T0296574

RIDERS IN GEOMETRY

RIDERS IN GEOMETRY

By

T. H. WARD HILL, M.A.

Senior Mathematical Master at Llandovery College
Late Scholar of Queen's College
Oxford

CAMBRIDGE
AT THE UNIVERSITY PRESS
1929

CAMBRIDGE
UNIVERSITY PRESS

University Printing House, Cambridge CB2 8BS, United Kingdom

Cambridge University Press is part of the University of Cambridge.

It furthers the University's mission by disseminating knowledge in the pursuit of education, learning and research at the highest international levels of excellence.

www.cambridge.org
Information on this title: www.cambridge.org/9781316611821

First published 1929
First paperback edition 2016

A catalogue record for this publication is available from the British Library

ISBN 978-1-316-61182-1 Paperback

PREFACE

Riders in general fall into two groups—those whose solution can be deduced in a more or less straightforward manner from the facts given or implied in the statement of the question; and those which can only be solved by a person possessing some mathematical sense or intuition.

We do not propose to consider the second kind. For the student of Matriculation or School Certificate standard their appearances in examination papers, happily, are rare.

This book does not profess to be a kind of key, or set of solutions, to a series of important riders; its object is rather to teach the student to think in an orderly manner, to fix his ideas at once upon the essential points in each question, and to enable him to note at once the relevant facts and deductions from them. The average student, when given a rider to solve, immediately commences trying this and trying that without any definite purpose. For the person of distinct mathematical ability that method has its advantages; but the author has found from experience that for the large majority of boys, working in this way without any ordered method, time is merely wasted; the boy becomes confused, loses interest, gets the idea that riders are beyond his powers and gives up all hope.

The illustrations have, for the most part, been taken from the "stock" riders. Once these have been acquired and understood, progress is easy. A few hints on the "form" of solutions are given in the introduction. No effort has been made to reproduce complete solutions in their finished form; that is left to the student. The aim has been to suggest how things are deduced, to link up the thought-processes of the mind.

Stress has been laid upon the fact that the enunciations of the various propositions in Geometry are really descriptions of the properties possessed by the figures under consideration.

Exercises have been added at the end of each section; each exercise will be found to deal primarily with the ideas suggested in the section. They have been selected with a view to giving the reader an opportunity of using his own powers of deduction on the lines laid down in the book. The Senate of the University of London have kindly granted permission for the use of questions set at their Matriculation Examination.

The order followed in reading the book may be varied at will. The chapter on Loci has been placed at the end because it serves, in that position, as a general revision of the whole book.

The author has again gratefully to acknowledge his indebtedness to Mr G. L. Parsons, M.A., of Merchant Taylors' School, who has read both the MS. and the proof-sheets and made many valuable suggestions.

T. H. W. H.

September 1929

CONTENTS

THE STRAIGHT LINE

THE STRAIGHT LINE

CHAPTER I

INTRODUCTION

Before commencing a rider read it through very carefully. It is a good plan to underline words or phrases that give rise to definite ideas, such as parallels, the angle BAC is bisected, $AB = AC$, two right angles..., and also what is required to be proved.

All the given facts must be used; the first thing to be done is to associate with each of these facts the properties connected with it. At the head of each chapter in what follows will be found the main properties discussed in that chapter.

The solution of riders consists in the main of eliminating the non-useful properties associated with each fact, and of linking up the remainder. Indications as to the proper sequence of the relevant facts invariably present themselves.

Special properties of any figures mentioned should be noted. Thus: a rectangle is a parallelogram with one of its angles a right angle; the diagonals of a rhombus bisect one another at right angles. These things often furnish a clue to the solution.

Some riders are with advantage worked "from both ends". In other words, some of the steps previous to that giving a final solution are retraced with a view to seeing whether they accord with the conclusions arrived at directly from the given facts. If that is so, the whole solution may then be built up. Examples of this method will be found on pp. 74, 105.

Great care must be taken in the drawing of diagrams. When given a triangle (without any other qualifications), do not draw an isosceles or an equilateral triangle, and be careful not to draw a parallelogram or a rectangle when you are simply given a quadrilateral. It is equally important that, when a figure is given any special characteristics, due attention is paid to them in drawing it. Thus if in a triangle $AB > AC$, or $AB = AC$, do not draw a triangle in which $AB < AC$, or even one in which AB is only slightly greater than AC. A warning, too, must be given against such practices as regarding things as equal because they look equal; assuming two straight lines perpendicular when they seem to cut at right angles. Of course many equalities are often suggested by the appearance of the figure, but such things must not be taken for granted unless they can be proved, and are proved. This does not mean that very accurate figures need be drawn with compasses and set-squares, etc.; a little care will always provide a satisfactory figure without the waste of any time.

It is a good plan to indicate equal lengths or angles in a figure by suitable marking: +, ‡ for lines, $\angle$o$\angle$o, $\angle$×$\angle$× or letters $\alpha\alpha$, $\beta\beta$ for angles, with a small c for common elements.

Proofs should be set out neatly; there should be only one step, however short it may be, to a line. References should be given wherever possible; these should not be made by numbers, but as far as possible by names, e.g.

> Three sides (in proving Congruence);
> Intercept theorem;
> Pythagoras' theorem;
> Rectangle theorem;
> Vertically opposite;
> Angle-sum;...

Finally, it is useful to remember that some riders (and also some constructions) have what may be called a family connection, and are conveniently grouped as such. We have, to give two instances:

The Kites,..., see pp. 12, 80 (q. 9), 107 (q. 11).

Riders in which we have to produce a median to twice its length,..., see pp. 14, 48.

CHAPTER II

PARALLELS

When we think of parallels only three main ideas need be exploited. If they are met by a transversal,

(i) *the alternate angles are equal;*

(ii) *the corresponding angles are equal;*

(iii) *the allied* angles are supplementary.*

There are three other possible ideas, but they are dependent upon particular conditions and can at once be isolated:

(iv) *parallels in connection with areas*—the idea of areas must necessarily be present. We deal with this in chapter VIII;

(v) *parallels in connection with mid-points of the sides of a triangle or trapezium*—for this see chapter V;

(vi) *parallels in connection with the ratio propositions*—this group is at once recognised by a direct reference to, or implication of, ratios.

If in a rider, when ideas (iv)–(vi) are not present, parallel straight lines are mentioned, one of the facts (i)–(iii) will have to be used, either in the direct or converse form.

The proof of the properties of a parallelogram is a typical example merely of writing down the consequences of the fact that the opposite sides of a parallelogram are parallel.

---

* We follow a recently suggested contraction for "interior angles on the same side of the cutting line".

Illustrative Riders

If the external bisector of an angle of a triangle is parallel to the opposite side, the triangle is isosceles.

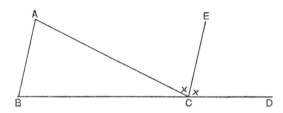

Ideas:

(i) Mark the equal angles ACE, ECD.

(ii) Keep in mind that an isosceles triangle in terms of angles (for parallels lead us to angles) means two equal angles.

(iii) AB and EC are parallel. We must use this.

We have
$$B\hat{A}C = A\hat{C}E \qquad \text{(alternate)},$$
$$A\hat{B}C = E\hat{C}D \qquad \text{(corresponding)}.$$

[Note that we have brought in the angles ACE, ECD about which we know something, namely that they are equal.]

So
$$B\hat{A}C = A\hat{B}C,$$

i.e.
$$AC = BC.$$

AD, the internal bisector of the $B\hat{A}C$ of a $\triangle ABC$, meets BC at D; a parallel through C to AD meets BA produced at E. Prove that $AC = AE$.

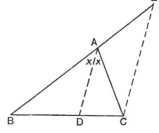

Ideas:

   (i) We have to prove
        $$AC = AE.$$

Translating this into terms of angles (because of parallels), we see that we have to prove

$$A\hat{C}E = A\hat{E}C.$$

   (ii) What use can we make of the fact that *AD* is parallel to *CE*? Keep $A\hat{C}E$ and $A\hat{E}C$ in mind.

We have $A\hat{C}E = D\hat{A}C$ (alternate),

                        $A\hat{E}C = B\hat{A}D$ (corresponding).

This must help us, as we are bringing $B\hat{A}D$ and $D\hat{A}C$ into the question.

   (iii) Use $B\hat{A}D = C\hat{A}D$

        and we have at once

$$A\hat{C}E = A\hat{E}C.$$

The solution may now be reconstructed.

Riders into which "allied" angles enter have a direct connection with one or two right angles, and the angle-sum theorem will frequently be necessary. The following is a typical example:

ABCDE is a regular pentagon. Prove that CE is parallel to AB.

(We shall here assume the properties
of a regular pentagon, i.e.

$$AB = BC = CD = DE = EA$$

and $\quad \hat{A} = \hat{B} = $ etc. $= 108°$;

and the angle-sum theorem see p. 39.)

Ideas:

    (i) To prove *CE* parallel to *AB*
we must have:

        (*a*) a pair of alternate angles equal;

or   (*b*) a pair of corresponding angles equal;

or   (*c*) a pair of allied angles supplementary.

    (ii) What do we know?

We have $E\hat{A}B = 108°$. Can we evaluate $A\hat{E}C$ and use (*c*)?
(To make use of (*a*) or (*b*) we should have to produce *CE*
or *AE*, and it would still be necessary to evaluate $A\hat{E}C$.)

    (iii) Of the angles at *E* we know $A\hat{E}D = 108°$. Can we
           evaluate $D\hat{E}C$?

Now $D\hat{E}C$ is an angle of a triangle—an isosceles triangle.

But $\qquad\qquad\qquad E\hat{D}C = 108°,$

so $\qquad\qquad\qquad D\hat{E}C = D\hat{C}E = \frac{1}{2}(180 - 108°)$
$$= 36°.$$
$$\therefore A\hat{E}C = 108° - 36°$$
$$= 72°,$$
so $\qquad\qquad E\hat{A}B + A\hat{E}C = 108° + 72°$
$$= 180°.$$
$$\therefore AB \text{ and } CE \text{ are parallel.}$$

Further examples having a bearing on this chapter are
worked on pp. 37, 96 and 102.

Exercise 1

1. Prove that straight lines which are parallel to the same straight line are parallel to one another.

2. If a straight line is perpendicular to one of a system of parallel straight lines, it is perpendicular to all the lines of the system.

3. If the opposite sides of a quadrilateral are parallel, its opposite angles are equal.

4. Similarly for a hexagon, or for any figure with an even number of sides.

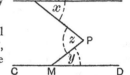

5. AB and CD are two parallel straight lines; L is any point in AB, M is any point in CD. Prove (see figure) that $z = x + y$.

6. Prove, similarly, in this figure that
$$B + D = A + C + E.$$
Give a generalised form of this rider.

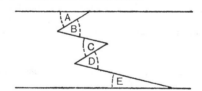

7. If two intersecting straight lines are respectively parallel to two other intersecting straight lines, prove that the angle between the first pair is equal, or supplementary, to that between the second pair.

8. Prove that in an isosceles triangle the external bisector of the vertical angle is parallel to the base.

9. In a $\triangle ABC$, $AB = AC$. DE is drawn parallel to BC to meet AB and AC respectively at D and E. Prove that $AD = AE$.

10. Straight lines are drawn through the vertices of a triangle parallel to the opposite sides. Prove that the triangle so formed is similar to the original triangle (i.e. that they are equiangular).

11. AD is the external bisector of the angle BAC of a $\triangle ABC$. CE is drawn parallel to AD to meet AB at E. Prove that $AC = AE$.

[You will have less difficulty with your figure if you make angle ACB obtuse.]

12. Two straight lines AB, CD intersect at O. If the bisectors of the adjacent angles AOC, COB meet a parallel through C to AB in P and Q, prove that $PC = OC = CQ$.

13. The internal bisectors of the angles B and C of a $\triangle ABC$ meet at O. A straight line is drawn through O parallel to BC and meets AB, AC at D and E respectively. Prove that

$$DE = BD + CE.$$

14. The equal angles B and C of an isosceles triangle ABC are bisected by lines BY, CX which meet the opposite sides in Y and X. Prove that XY is parallel to BC and also that

$$BX = XY = YC.$$

CHAPTER III

CONGRUENCE

To prove the congruence of two triangles it is necessary that the following elements of the one triangle be equal to the corresponding elements of the other:

(i) *two sides and the angle included by them;*

or (ii) *the three sides;*

or (iii) *two angles and a side.*

There is one special case:

(iv) *if the two triangles are right angled, it is sufficient that any two sides of the one are equal to the corresponding sides of the other.*

The procedure to be followed in the proving of congruence is as follows. Examine the two triangles to see if the conditions of any of the above cases hold. If so, the proof may at once be constructed. Notice particularly whether the triangles have a common side or a common angle, or whether vertically opposite angles appear.

Frequently only two equal elements are given. If *two angles* of one triangle are equal to the corresponding angles of the other triangle, look out for a pair of equal sides (or a common side), or for a means of proving a pair of sides equal (e.g. the isosceles triangle theorem).

If *two sides* of one triangle are equal to the corresponding sides of the other, examine the figure to see if it is at all possible to prove their included angles equal. If not, try the third sides.

If *one side and one angle* of a triangle are equal to the corresponding side and angle of the other then, if the side and the angle face each other and the equal angles are acute, congruence can only be proved by first establishing the equality of a second angle of the one triangle with the corresponding angle of the other. If the given angle is a right angle or an obtuse angle, congruence may also be proved by means of another pair of equal corresponding sides. (Cf. the "Ambiguous Case" in the drawing of triangles.)

It must not be supposed that nothing can be said of two triangles having two sides and a facing angle (a non-included angle) of the one equal to the corresponding elements of the other. The triangles *may* be congruent (if certain other conditions are satisfied); if not, then the other pair of facing angles are supplementary.

This point of an angle and facing side is of such importance that we give an example of false congruence to emphasise it.

P is any point (except the mid-point) in the base BC of an isosceles triangle ABC.

Consider the △s *ABP, ACP.*

In them we have

$\begin{cases} AB = AC & \text{(equal sides),} \\ AP \text{ is common,} \\ A\hat{B}P = A\hat{C}P & \text{(angles at the base).} \end{cases}$

But the triangles are *not* congruent,

e.g. $BP \neq PC.$

We have made use of an acute angle and facing side, and then another side.

We can say, however, that the angles *APB, APC* are supplementary; of course this is otherwise obvious.

If the side and angle do not face, then the other arm of the angle may be considered, or else a second angle; but care must be taken that the equal sides face corresponding angles, and vice versa.

Equal straight lines and equal angles

It is useful to note in what *possible* connections equal straight lines or equal angles may be met. They are:

Equal straight lines.

Corresponding sides of congruent triangles.
Sides opposite equal angles of an isosceles triangle.
Opposite sides of a parallelogram, or diagonals bisected.
Equal intercepts (intercept and mid-point theorems).
Equal chords in a circle.
Equal tangents from an external point to a circle.

Equal angles.

Vertically opposite angles.
Corresponding angles of congruent triangles.
Equal base angles of an isosceles triangle.
Corresponding or alternate angles formed by a transversal to parallel straight lines.
Opposite angles of parallelograms.
Those arising from the angle, arc, and tangent properties of a circle.

Illustrative Riders

ABCD is a kite (i.e. $AB = AD$ *and* $BC = CD$). *Prove that AC bisects BD at right angles.*

Ideas:

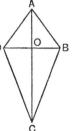

(i) What are we given? Only that
$AB = AD$ and $BC = CD$.

(ii) What do we want to prove? That
$DO = OB$ and $A\hat{O}D = 1$ rt. $\angle$.

But if $A\hat{O}D = 1$ rt. $\angle$, so does $A\hat{O}B$.

We have therefore to prove

$$DO = OB \text{ and } A\hat{O}D = A\hat{O}B.$$

(iii) We are thus led to consider the congruence of
△s *AOD*, *AOB*.

In them we have $\begin{cases} AD = AB \\ AO \text{ common}; \end{cases}$ (given)

two sides, so we must prove either

(*a*) *DO* = *OB*—but this begs the question,

or (*b*) the included angles *OAD*, *OAB* equal.

We must therefore prove

$$O\hat{A}D = O\hat{A}B.$$

(iv) They are not in the same triangle, so the isosceles
triangle theorem is of no use. In fact we may rule out
all the possibilities given on p. 12 except congruence
of triangles. We must therefore look for two triangles
containing these angles and try to prove them congruent.

(v) Triangles *ADC*, *ABC* are the only ones possible, and
they are congruent, for

$$\begin{cases} AD = AB & \text{(given)} \\ CD = BC & \text{(given)} \\ AC \text{ is common.} \end{cases}$$

The proof may now be reconstructed.

If the bisector of the vertical angle of a triangle bisects the base, the triangle is isosceles.

Ideas:

(i) The congruence of the △s *AOB*, *AOC* seems to be suggested.

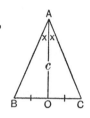

In them we have

$$\begin{cases} BO = OC & \text{(given)} \\ AO \text{ is common} \\ B\hat{A}O = C\hat{A}O & \text{(given)}. \end{cases}$$

But the first and third data give a side and facing angle, so we must prove another pair of angles equal. This, however, begs the question. We must therefore think of something else.

(ii) *AO* is a median, so let us try the median construction (i.e. produce *AO* to *D* making *OD = AO* and join *CD*).

N.B. This construction is always worth considering when a median is mentioned.

(iii) The △s *AOB*, *COD* are congruent (two sides and their included angle).

(iv) How can we make use of

$$B\hat{A}O = C\hat{A}O?$$

From △s *AOB*, *COD* we have

$$B\hat{A}O = C\hat{D}O,$$

so

$$C\hat{A}O = C\hat{D}O,$$

i.e.

$$AC = CD,$$

the only possible deduction.

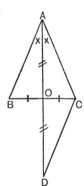

(v) Now we wish to prove

$$AB = AC.$$

But

$$AB = CD; \text{ hence, etc.}$$

Equilateral triangles BCX, CAY, ABZ are described out-wards on the sides BC, CA, AB respec-tively of a triangle ABC. Prove that

$$AX = BY = CZ.$$

Ideas:

Let us first take $AX = CZ$.

(i) $AX = CZ$ cannot mean:

    (*a*) an isosceles triangle;

    (*b*) opposite sides of a paral-lelogram;

    (*c*) equal intercepts;

so we try congruence.

(ii) What are we given? Only that equilateral triangles are drawn on the sides of the triangle. Hence

$$AB = BZ = ZA; \quad BC = CX = XB; \quad AC = CY = YA.$$

(iii) We must now find triangles which bring in AX, CZ and some of the quantities in (ii).

We have in $\triangle$s ABX, CBZ (why not $\triangle$s ABX, AZC?)

$$\left.\begin{matrix} AB = BZ \\ BX = BC \end{matrix}\right\} \text{ sides of equilateral triangles.}$$

We cannot use $AX = CZ$ (third sides), so we must examine the included angles ABX, CBZ.

(iv) These angles have $A\hat{B}C$ in common. Take it away from each. It remains to be proved that

$$C\hat{B}X = A\hat{B}Z.$$

But each of these angles is 60° (equilateral triangles). The proof may now be reconstructed.

*BAC is any angle. P and M are points on AB, and Q and N
are points on AC, such that*

$AP = AQ$ and $AM = AN$.

*If PN, QM cut at S, prove that
AS bisects the angle BÂC.*

Ideas:

    (i) As before, we fall back
       upon congruence. (Try
       the other possibilities for yourself.)

    (ii) $\triangle$s *AMS, ANS* or $\triangle$s *APS, AQS* are indicated.

    (iii) Let us first consider $\triangle$s *AMS, ANS*.

In them we have $\begin{cases} AM = AN & \text{(given)} \\ AS \text{ is common.} \end{cases}$

We cannot hope to use the included angles, as they are to
be proved equal.

We have therefore to fall back upon proving

$$MS = NS \text{ (third sides)}$$

or else $M\hat{S}A = N\hat{S}A$ and $A\hat{M}S = A\hat{N}S$ (two angles).

A little reflection shows us that we can scarcely expect to
be able to prove $M\hat{S}A = N\hat{S}A$; for these angles only come
into the question by joining AS and nothing from p. 12 can
help us.

So let us try to prove $MS = NS$.

    (iv) This means proving $\triangle$s *MSP, NSQ* congruent.
In them we have

$\begin{cases} MP = NQ & \text{(differences of equal lengths)} \\ M\hat{S}P = N\hat{S}Q & \text{(vertically opposite).} \end{cases}$

Here we have a pair of sides and facing angles. We must
therefore prove another pair of angles equal, say $M\hat{P}S = N\hat{Q}S$
($P\hat{M}S = Q\hat{N}S$ would do equally well).

(v) Now $M\hat{P}S$, $N\hat{Q}S$ belong to the $\triangle$s APN, AQM; and in these triangles

$$\begin{cases} AP = AQ & \text{(given)} \\ AN = AM & \text{(given)} \\ \text{incl. } P\hat{A}Q \text{ is common.} \end{cases}$$

So the triangles are congruent; hence, etc.

Ex. Obtain the same result, starting from $\triangle$s APS, AQS.

Exercise 2

1. Prove that points which are equidistant from the extremities of the base of an isosceles triangle are also equidistant from the vertex.

2. $ABCD$ is a square. P, Q, R, S are the mid-points of AB, BC, CD, DA respectively. Prove that

   (i) $AQ = DQ$;   (ii) $PQ = QR$;   (iii) $AQ = CP$.

3. Prove that all points on the bisectors of an angle are equidistant from the arms of the angle.

4. State and prove the converse of question 3.

5. From any point on the bisector of an angle parallels are drawn to the arms of the angle. Prove that the quadrilateral so formed has all its sides equal.

6. Prove that each diagonal of a rhombus bisects the angles through which it passes.

7. If the four sides of a quadrilateral are equal, prove that its diagonals bisect each other at right angles.

8. Prove that the common chord of two intersecting circles is bisected at right angles by the straight line joining their centres.

[The radii of a circle are equal.]

9. P is a point within an $A\hat{O}B$. OP is produced to Q so that $PQ = OP$. Through Q QR is drawn parallel to OA, meeting OB in R; RP produced meets OA at T. Prove that RT is bisected at P.

Hence devise a construction for drawing a straight line through a given point within an angle, and terminated by the arms of the angle, which shall be bisected at that point.

10. $ABCD$, $PQRS$ are two quadrilaterals in which $AB = PQ$, $BC = QR$, $CD = RS$, $DA = SP$ and $\hat{A} = \hat{P}$. Prove that the quadrilaterals are equal in all respects.

11. Two quadrilaterals have three sides of the one equal to the corresponding sides of the other and their diagonals equal to each other. Prove that their fourth sides are also equal.

12. OA and OB are two straight lines intersecting at O. OX is drawn perpendicular to OA and equal to OA, and OY is drawn perpendicular to OB and equal to OB, OX and OY both being drawn outwards from the angle AOB. Prove that
$$AY = BX.$$

13. Two isosceles triangles AOB, COD have a common vertex O and equal vertical angles. Show that $AC = BD$.

14. ABC is an equilateral triangle; BC, CA, AB are produced for equal lengths to P, Q, R respectively. Show that the triangle PQR is also equilateral.

15. ABC, DBC are two triangles on the same base BC and on the same side of it, and $\hat{A} = \hat{D}$, $A\hat{B}C = D\hat{C}B$. Prove that the straight line joining the point of intersection of AC and BD to the mid-point of BC is perpendicular to BC.

16. AB, $A'B'$ are two equal straight lines; the perpendicular bisectors of AA', BB' meet at O. Show that $A\hat{O}B = A'\hat{O}B'$.

17. ABC, $A'B'C'$ are two congruent triangles. Prove that the perpendicular bisectors of AA', BB', CC' meet at a point. What happens if the sides of $\triangle A'B'C'$ are respectively parallel to those of $\triangle ABC$?

18. Prove that two triangles are congruent if two sides and a median of one are respectively equal to the corresponding sides and median of the other. [Two cases.]

19. P is the point where the perpendicular bisector of a side BC of a $\triangle ABC$ meets the bisector of $B\hat{A}C$. PL, PM are drawn perpendicular to AB, AC respectively (produced if necessary). Prove that $AL = AM$ and $BL = CM$.

20. From CA, a diagonal of a square $ABCD$, CE is cut off equal to CB, and EF is drawn perpendicular to AC, meeting AB in F. Prove that $AE = EF = BF$.
[Join CF.]

21. On the same base BC and on the same side of it two triangles ABC, DBC are drawn in which $AB = AC$ and $DB = DC$; prove that AD, when produced, bisects BC.

22. A quadrilateral $ABCD$ has the adjacent sides AB, AD equal, and also the opposite angles B, D equal. Prove that $BC = CD$ and that its diagonals are perpendicular to each other.
[Equal adjacent sides?]

23. $ABCD$ is a square. Points P, Q are taken in the sides BA, BC respectively such that $BP = BQ$; AQ, CP cut at O. Prove that $AO = OC$.

24. $ABCD$ is a quadrilateral in which the diagonal AC bisects $\hat{A}$ and $CD = CB$. If $AD \neq AB$, prove that the angles B and D are supplementary.

25. In a quadrilateral $ABCD$, $AB = DC$, $\hat{B} = \hat{D}$. Prove that it is not necessarily a parallelogram.

CHAPTER IV

PARALLELOGRAMS

A parallelogram is *defined* as a quadrilateral whose opposite sides are parallel.

The main propositions on parallelograms establish their *properties*. So, when a parallelogram is mentioned, we must at once bring to mind:

(i) *its opposite sides are parallel;*

(ii) *its opposite sides are equal;*

(iii) *its opposite angles are equal;*

(iv) *each diagonal bisects the parallelogram;*

(v) *the diagonals bisect each other;*

(vi) *the area propositions.*

In connection with (iii) it is useful also to remember that adjacent angles of a parallelogram are supplementary.

(iv) is chiefly of use in area questions. These are dealt with, together with (vi), in a later chapter.

Exercises 1–5 on p. 26 will be found to follow at once from the above properties of a parallelogram.

Illustrative Riders

To prove that a parallelogram whose diagonals are equal must be a rectangle.

Ideas:

(i) What have we to prove?— that one of the angles of the parallelogram is a right angle (definition of a rectangle).

This means that all the angles must be right angles. We have therefore to prove that

$$D\hat{A}B = C\hat{B}A \quad \text{(say)}.$$

(ii) What are we given?

    (*a*) a parm. *ABCD*.

      ∴ opposite sides, angles equal, etc.

    (*b*) $AC = BD$.

(iii) Referring to p. 12 we see that in order to prove $D\hat{A}B = C\hat{B}A$ we shall have to prove a pair of triangles congruent.

△s *DAB*, *CBA* suggest themselves, for in them

$$\begin{cases} BD = AC & \text{(given)} \\ AB \text{ is common} \\ AD = BC & \text{(opposite sides of parm.).} \end{cases}$$

      ∴ the triangles are congruent,

so $$D\hat{A}B = C\hat{B}A,$$

but $D\hat{A}B + C\hat{B}A = 2$ rt. ∠s (*AD*, *CB* are ∥); hence, etc.

Note. Since diagonals are mentioned, there is another line of attack:

    (i) $AO = OC$, $DO = OB$ and $AC = BD$,

so $$AO = OB.$$

    (ii) Now use the angle-sum theorem, or prove directly that

$$O\hat{A}B + O\hat{A}D = O\hat{D}C + O\hat{D}A, \text{ etc.}$$

ABCD, APCQ are two parallelograms having a common diagonal AC. Prove that PBQD is a parallelogram.

Ideas:

- (i) If *PBQD* is a parm., we must either prove:
  - (*a*) its opposite sides parallel;
  - (*b*) its opposite sides equal;
  - (*c*) its opposite angles equal;
  - (*d*) a pair of sides equal and parallel;
  - (*e*) its diagonals bisect each other.

- (ii) (*a*), (*c*) and (*d*) each entail a knowledge of the angles of *PBQD*. But as the sides of *PBQD* are "joins" of points, we cannot expect to know anything about its angles. (It would be different if the points were on the circumference of a circle, for instance; but here we have no direct indication of, or help with regard to, the angles.)

So it is not worth commencing, at any rate, with (*a*), (*c*) and (*d*).

(iii) Let us consider (*b*); e.g. can we prove $PD = BQ$?

Referring to p. 12 we see that this will most probably mean the proving of the congruence of two triangles. What are we given?—two parms. *ABCD, APCQ.* We therefore have opposite sides equal, etc.

We note that in the △s *ADP, BCQ* (to which *PD, BQ* belong)

$$\begin{cases} AD = BC & \text{(opposite sides of parm.)} \\ AP = CQ & \text{(opposite sides of parm.).} \end{cases}$$

We must therefore try to prove the included angles equal,
i.e. $D\hat{A}P = Q\hat{C}B$ (why the angles?).

But $D\hat{A}C = A\hat{C}B$ (alternate),

                        $P\hat{A}C = A\hat{C}Q$ (alternate);

hence, adding, etc.

> (iv) We have not yet ruled out (i) (e); i.e. to prove that
> BD and PQ bisect each other.

Now BD is a diagonal of the parm. $ABCD$.

So the mid-point of BD is the mid-point of AC.

Similarly the mid-point of PQ is the mid-point of AC, from
parm. $APCQ$.

We have therefore another demonstration of this rider.

Equal and parallel straight lines

Many riders may be solved by the application of the fact that if the ends of equal and parallel straight lines are joined towards the same parts the joins themselves are equal and parallel; in other words, that the figure so formed is a parallelogram.

ABCD is a parallelogram and E and F are the mid-points of the opposite sides BC and AD. Prove that AECF is a parallelogram.

Ideas:

(i) What do we know about *AECF?—AF* and *EC* are ∥.

(ii) We must therefore either prove that *AE* and *FC* are ∥, or that $AF = EC$.

(iii) We must use the fact that E and F are mid-points.

Now $AD = BC$, and F, E are their mid-points;

so $AF = EC,$

i.e.   *AF* and *EC* are equal and parallel,

i.e.   *AECF* is a parallelogram.

D is the mid-point of the side BC of a triangle ABC. AM is drawn equal and parallel to BD. Prove that DM bisects AC, say at K; and that $KD = \frac{1}{2}AB$.

Ideas:

(i) AM and BD are equal and parallel; therefore AB and DM are equal and parallel.

(ii) How can we bring in AC, and in particular the mid-point of AC? We must also use the fact that $BD = DC$—we have not done so yet. This we can do by saying that

AM and DC are equal and parallel,

so $ADCM$ is a parallelogram.

(iii) AC, DM are diagonals of this parallelogram, and we are discussing their point of intersection, K.

We have at once that AC and DM bisect each other at K.

So
$$KD = \tfrac{1}{2}DM$$
$$= \tfrac{1}{2}AB$$

Exercise 3

1. Any straight line drawn through the point of intersection of the diagonals of a parallelogram and terminated by a pair of parallel sides is bisected at that point.

2. The straight line of question 1 bisects the area of the parallelogram.

3. Draw straight lines bisecting the area of a parallelogram to fulfil the following conditions:
    (i) to pass through a given point;
    (ii) to be parallel to a given straight line;
    (iii) to be perpendicular to a side of the parallelogram.

4. If the diagonals of a quadrilateral bisect each other, the quadrilateral is a parallelogram.

5. If the opposite sides of a quadrilateral are equal, the quadrilateral is a parallelogram.

6. Through the vertices of a triangle ABC equal and parallel straight lines AA', BB', CC' are drawn in the same sense. Prove that the triangles ABC, $A'B'C'$ are congruent. Give a generalisation of this rider.

7. If two pairs of opposite sides of a hexagon are equal and parallel, show that the third pair are also equal and parallel.

8. $ABCD$, $ABEF$ are two parallelograms having a common side AB. If CE, DF be joined, show that the figure $CDFE$ is a parallelogram.

9. $ABCD$ is a parallelogram; L and M are the mid-points of AB and DC respectively. Prove that AM, BM, CL, DL form a parallelogram.

10. Deduce from the fact that the diagonals of a rectangle are equal that the straight line joining the vertex of the right angle to the mid-point of the hypotenuse of a right-angled triangle is equal to half the hypotenuse.

11. Prove that two parallelograms are congruent if two adjacent sides of the one are equal to two adjacent sides of the other, and the angles included by those sides are supplementary.

12. $ABCD$ is a parallelogram; E is a point on the diagonal AC. Through E parallels are drawn to AD and AB terminated by AB, CD, AD, BC at P, Q, R, S respectively. Prove that the parallelograms $PBSE$, $DQER$ are equal in area.

[Each diagonal bisects a parallelogram.]

13. Prove that any point on the perpendicular bisector of the cross-bar of a pair of equal Rugby Football goal posts is equidistant from the tops, and from the bases, of the posts.

14. $ABCD$ is a rhombus, and BA is produced to E so that $AE = BA$ and EC is joined. Prove that

(i) EC bisects AD;

(ii) EC divides the diagonal BD in the ratio 2 : 1.

15. Prove that if two sets of straight railway lines of equal gauge cross one another, they form a rhombus.

16. In a trapezium $ABCD$ the longest side AD is double each of the other sides (which are equal). If O is the middle point of AD, show that the triangle OBC is equilateral.

17. $ABCD$ is a square and points P, Q, R, S are taken in the sides AB, BC, CD, DA respectively, such that $AP = CR$ and $BQ = DS$. Show that PR and QS intersect at the centre of the square.

[Consider the intersection of PR with a diagonal.]

CHAPTER V

THE MID-POINT AND THE INTERCEPT PROPOSITIONS

A very important class of riders follows from the group:

(i) *if a number of parallel straight lines make equal intercepts on one transversal, they will make equal intercepts on any transversal* (a particular case arises when the straight lines are given as equidistant);

(ii) *the straight line drawn through the mid-point of one side of a triangle parallel to another side bisects the third side;*

(iii) *the straight line joining the mid-points of two sides of a triangle is parallel to the third side and equal to a half of it.*

The clues for this class are:

(i) mention of more than two parallel straight lines, coupled with lengths of straight lines intersecting them;

(ii) one mid-point and a parallel;

(iii) two mid-points of *adjacent sides* of a figure.

(i) is of fundamental importance in the geometrical theory of ratios. (ii) and (iii) are sometimes met in connection with areas, and in this connection are dealt with in a later chapter on areas.

Illustrative Riders

Prove that if the mid-points of the sides of a quadrilateral are joined in order, the joins form a parallelogram.

Ideas:

(i) We are given straight lines joining mid-points of adjacent sides of a figure. Take SR; this suggests forming the triangle ADC, so join AC.

(ii) We have at once that

$SR = \frac{1}{2}AC$ and is $\parallel$ to AC,

and similarly

$PQ = \frac{1}{2}AC$ and is $\parallel$ to AC, from $\triangle\ ABC$.

So $\qquad SR$ and PQ are equal and parallel,

i.e. $\qquad\qquad PQRS$ is a parallelogram.

Or, we might have said

(iii) $\qquad SR$ and PQ are parallel.

Similarly PS and QR are parallel, by joining BD.

So $\qquad\qquad PQRS$ is a parallelogram.

D and E are the mid-points of the sides AB, AC of a triangle ABC; F is any point in BC. Prove that AF is bisected by DE.

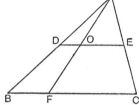

Ideas:

(i)   *D* and *E* are the mid-points of adjacent sides of the triangle *ABC*.

We at once associate with this that *DE* is ∥ to *BC*.

(ii) We want to prove that *O* is the mid-point of *AF*.

We think of the mid-point propositions. The third one given above clearly does not apply, for it pre-supposes that two mid-points are given. The conditions of the second one, however, hold for *D* is the mid-point of *AB*, and *DO* we have just proved is parallel to *BC*.

Hence *O* is the mid-point of *AF*.

Aliter by the Intercept Theorem.

(i)   First prove *DE* is parallel to *BC*. *AF* is another transversal through *A*.

(ii) Draw a parallel through *A* to *BC* and we can at once apply the intercept theorem.

The internal bisector of the angle BAC of a triangle ABC meets BC in D. CL is drawn perpendicular to AD; M is the mid-point of BC. Prove that

I. *LM is parallel to AB.*

II. $LM = \frac{1}{2}(AB \sim AC)$.

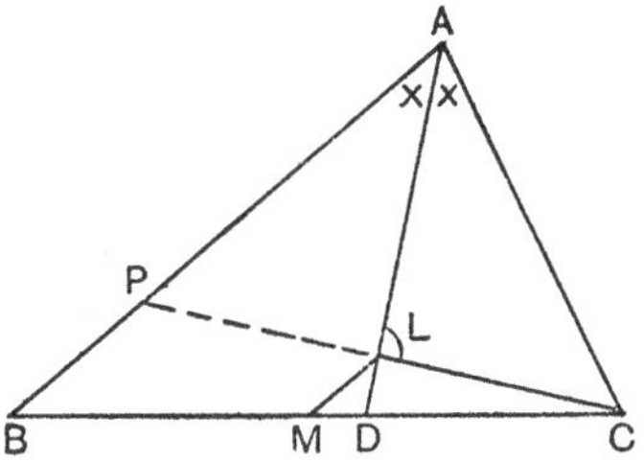

Ideas:

I. (i) We have to prove that *LM* is parallel to *AB*; but *M* is the mid-point of *BC*.

This suggests that *L* has to be the mid-point of—what?
Let us produce *CL* to meet *AB* at *P*.
Can we prove that *L* is the mid-point of *CP*?

(ii) What have we not yet used?

(a) $P\hat{A}L = C\hat{A}L$,

(b) $A\hat{L}C = $ 1 rt. $\angle$, and $\therefore = A\hat{L}P$.

We see then, adding to (a) and (b) that *AL* is common to both triangles, that the △s *ALP*, *ALC* are congruent;

whence $\qquad\qquad CL = LP$,

and so $\qquad LM$ is parallel to *BP*, i.e. to *AB*.

II. It follows that $LM = \frac{1}{2}BP$; [so *BP* must be proved equal to $AB \sim AC$].

But $\qquad\qquad BP = AB - AP$
$$= AB - AC.$$

Exercise 4

1. Prove that the straight lines joining the mid-points of the sides of a triangle divide the triangle into four congruent triangles.

2. Establish the truth of the rider on p. 28 for a skew quadrilateral, that is for a non-planar quadrilateral.

3. Prove that the straight lines joining the mid-points of opposite sides of a quadrilateral bisect one another.

4. Prove that the straight lines joining the mid-points of opposite sides of a kite are equal.

5. If the diagonals of a quadrilateral are equal, prove that the joins of the mid-points of opposite sides bisect each other at right angles.

6. $ABCD$ is a parallelogram; E and F are the mid-points of BC and AD. Show that AE and FC trisect the diagonal BD (see p. 24).

7. D and E are the mid-points of the sides AC, AB of a triangle ABC; CD is produced to H making $DH = CD$ and BE to K making $EK = BE$. Prove that HAK is a straight line, and parallel to BC.

8. E and F are the mid-points of the sides AC, AB of a triangle ABC. BE and CF meet in G. Produce AG to H so that $AG = GH$. Prove that $BGCH$ is a parallelogram, and deduce that the medians of a triangle are concurrent.

9. P and Q are the mid-points of the oblique sides DA, BC of a trapezium $ABCD$. By joining P and Q to the mid-point of either diagonal of the trapezium, show that PQ is parallel to AB and DC, and equal to $\frac{1}{2}(AB + DC)$.

10. Prove that the straight line joining the mid-points of the diagonals of a trapezium is equal to half the difference of the parallel sides.

11. O is the mid-point of a straight line AB. Prove that the sum of the perpendiculars from A and B upon any straight line which does not pass between A and B is equal to twice the perpendicular from O upon the same straight line.

12. If the straight line (see question 11) passes between A and B, then the difference of the perpendiculars is equal to twice the perpendicular from O.

13. Perpendiculars p_1, p_2, p_3, p_4 are drawn from the vertices, in order, of a parallelogram to any straight line which lies completely outside the parallelogram. Prove that

$$p_1 + p_3 = p_2 + p_4.$$

14. The figure shows a right section of a rectangular box $LMCN$ resting inside another one $ABCD$.

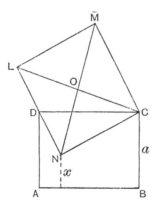

If $BC = a$, $ND = DL$, and the distance of N from AB is x, find in terms of a and x the distances of L, O and M from AB.

15. B', C' are the mid-points of the sides AC, AB of a triangle ABC; AD is the perpendicular from A to BC. Prove that the triangles $AB'C'$, $DB'C'$ are congruent.

16. In the figure the equal rods *AB*, *BC*, *CD*, *DE* are hinged at *B*, *C* and *D*. Show that if *A*, *C*, *E* are in a straight line, then so are the mid-points of the rods.

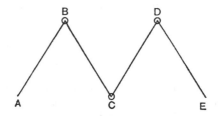

17. *P*, *U*, *R*, *T* are the mid-points of the sides *AB*, *BC*, *CD*, *DA* respectively of a quadrilateral *ABCD*; *Q* and *S* are the mid-points of the diagonals *AC* and *BD*. Prove that

(i) *PQRS* is a parallelogram;

(ii) *PR*, *SQ* and *TU* bisect one another and so are concurrent.

18. *ABCD* is a parallelogram; *AB* is produced to *E* making *BE* = *AB*, and *AD* to *F* making *DF* = *AD*. Prove that *BF* and *DE* meet on *AC*.

CHAPTER VI

THE ANGLE-SUM THEOREM

Right Angles

When **two right angles** (or multiples of two right angles) are mentioned, there are four possible ideas:

(i) *angles at a point;*

(ii) *allied angles and parallels;*

(iii) *the angle-sum property of a triangle, and its extensions to quadrilaterals and polygons;*

(iv) *the opposite angles of a cyclic quadrilateral.*

(ii) and (iv) are easily isolated, as the ideas of parallels, and of a circle, respectively, must enter the question. A useful variation of (iii) is that if a side of a triangle is produced the exterior angle so formed is equal to the sum of the interior opposite angles. This is frequently of value when dealing with angle questions arising from polygons.

Single right angles may be associated with either of the above ideas when *bisection* is mentioned or implied; the other possible cases are:

(v) *Pythagoras' Theorem and its converse;*

(vi) *the angle between a tangent and the radius at its point of contact;*

(vii) *the straight line joining the centre of a circle to the mid-point of a chord is perpendicular to the chord;*

(viii) *the angle in a semicircle.*

(v) will introduce squares on sides; (vi) and (vii) can be at once isolated; and (viii) will require a diameter of a circle.

Illustrative Riders

BO and CO are the bisectors of the angles B and C of a triangle ABC. Prove that $B\hat{O}C = 90° + \frac{1}{2}\hat{A}$.

Ideas:

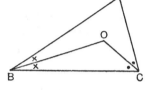

(i) The mention of a right angle and $\frac{1}{2}\hat{A}$ (-bisection-) suggests the angle-sum property.

(ii) Using this, we have
$$B\hat{O}C = 180° - O\hat{B}C - O\hat{C}B.$$

(iii) But we are given that BO, CO are the bisectors of $\hat{B}$ and $\hat{C}$; we must use this, so we have
$$B\hat{O}C = 180° - \tfrac{1}{2}\hat{B} - \tfrac{1}{2}\hat{C}.$$

(iv) The required result contains $\hat{A}$, but not $\hat{B}$ and $\hat{C}$. How can we connect them?

Well, $\hat{A} + \hat{B} + \hat{C} = 180°$,

so $\tfrac{1}{2}\hat{B} + \tfrac{1}{2}\hat{C} = 90° - \tfrac{1}{2}\hat{A}$,

whence $B\hat{O}C = 180° - (90° - \tfrac{1}{2}\hat{A}) = 90° + \tfrac{1}{2}\hat{A}$.

A transversal meets two parallel straight lines. Show that the bisectors of a pair of allied angles are at right angles.

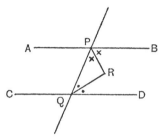

Ideas:

(i) Parallels suggest:
  (*a*) alternate angles equal;
  (*b*) corresponding angles equal;
  (*c*) allied angles supplementary.

(ii) (*c*) is the only fact which suggests a right angle.

(iii) Making use of bisectors, we have therefore

$$R\hat{P}Q + R\hat{Q}P = 90°.$$

(iv) Using the angle-sum theorem, we have at once

$$P\hat{R}Q = 90°.$$

If the opposite angles of a quadrilateral are equal, its opposite sides are parallel (i.e. *it is a parallelogram*).

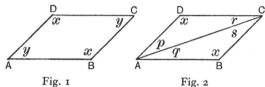

Fig. 1   Fig. 2

Ideas:

    (i) Mark the equal angles

$$\hat{B} = \hat{D} = x,$$
$$\hat{A} = \hat{C} = y \text{ (say).}$$

    (ii) One's first impulse is to draw a diagonal. Join AC. We have then, in the $\triangle$s ADC, ABC

$$\begin{cases} \hat{D} = \hat{B} & \text{(given)} \\ AC \text{ is common,} \end{cases}$$

i.e. a facing angle and side. So we must look for another pair of equal angles. But this would beg parallelism (why?). The congruence of the triangles is therefore ruled out.

    (iii) Our facts are concerned solely with angles; congruence has been ruled out, likewise equal adjacent angles in a triangle, so we are led to try the angle-sum property.

We may now proceed in two ways:

    (*a*) Taking the quadrilateral as a whole (Fig. 1) we have

$$x + y + x + y = 4 \text{ rt. } \angle\text{s,}$$

so $x + y = 2 \text{ rt. } \angle\text{s,}$

i.e. the opposite sides are parallel (allied angles);

or   (*b*) Using our diagonal (Fig. 2) we have

$$x + p + r = 2 \text{ rt. } \angle\text{s} = x + q + s,$$

so $p + r = q + s,$

but $p + q = r + s$ (given),

whence $q = r$, and AD and BC are $\parallel$, etc.

The calculation of the sides and angles
of a regular polygon

The data at our disposal are:

(i) *the sum of the interior angles + 4 right angles = 2n right
angles, where n is the number of sides;*

(ii) *if the sides are produced in order the sum of the exterior
angles so formed is 4 right angles.*

It will be found that the second of these expressions is the
more useful for numerical calculations. The methods to be
followed will be indicated by two examples.

*How many degrees are there in each interior angle of a regular
hexagon?*

$$\text{Each exterior angle} = \frac{360°}{6} = 60°.$$

$$\therefore \text{ each interior angle} = 180° - 60°$$

$$= 120°.$$

*How many sides has a regular polygon, each of whose interior
angles contains 140°?*

$$\text{Each exterior angle} = 180° - 140°$$

$$= 40°.$$

$$\therefore \text{ number of sides} = \text{number of exterior angles}$$

$$= \frac{360}{40}$$

$$= 9.$$

ABCDEF and ABXYZ are a regular hexagon and a regular
pentagon on the same base AB and
on the same side of it. YX produced
meets BC at P. Prove that

$$CX = PX.$$

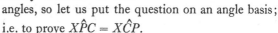

Ideas:

(i) Our knowledge of hexagons and pentagons is chiefly confined to their angles, so let us put the question on an angle basis; i.e. to prove $X\hat{P}C = X\hat{C}P$.

(ii) What do we know?
Each interior angle of the hexagon $= 120°$.
Each interior angle of the pentagon $= 108°$.
All their sides are equal.

(iii) How can we express $X\hat{P}C$? Not in terms of the angles of $\triangle XPC$, for that brings in $X\hat{C}P$ and $C\hat{X}P$ which we also want to find. So using the exterior angle property we have

$$X\hat{P}C = B\hat{X}P + P\hat{B}X$$
$$= (180° - 108°) + (120° - 108°) = 84°.$$

(iv) Now for $X\hat{C}P$. It is not exterior to any triangle in our figure but it is an angle of the $\triangle BCX$ in which $BC = BX$.

So $\qquad X\hat{C}P = C\hat{X}B = \tfrac{1}{2}(180° - C\hat{B}X)$
$$= \tfrac{1}{2}\{180° - (120° - 108°)\}$$
$$= \tfrac{1}{2}(168°) = 84°,$$

whence $\qquad CX = PX.$

Exercise 5

1. In a triangle ABC, BO bisects $A\hat{B}C$, and CO bisects $A\hat{C}B$ externally. Prove that $B\hat{O}C = \frac{1}{2}\hat{A}$.

2. AB and CD are parallel straight lines and a transversal meets them at P and Q respectively. Show that the bisectors of the four interior angles at P and Q form a rectangle, one of whose diagonals is parallel to AB and CD.

3. (a) How many degrees are there in each interior angle of a regular pentagon, heptagon, decagon, 18-sided figure?

(b) How many sides have regular polygons whose interior angles each measure $108°$, $135°$, $150°$, $172°$?

4. Prove that the bisectors of the angles of a parallelogram form a rectangle.

5. Prove that the bisectors of the angles of a quadrilateral form a quadrilateral whose opposite angles are supplementary. What does this quadrilateral become when the given quadrilateral is (a) a rectangle, (b) a square, (c) a rhombus, (d) a kite?

6. If two straight lines are respectively perpendicular to two other straight lines, prove that the angle between the first pair is equal or supplementary to that between the other pair.

7. $ABCDE$ is a regular pentagon. Prove that CE is parallel to AB. Extend this result to any regular polygon having an odd number of sides.

8. The sides (in order) of a pentagon inscribed in a circle subtend angles of $40°$, $50°$, $60°$, $100°$, $110°$ at the centre of the circle. Find the number of degrees in each angle of the pentagon.

9. In a quadrilateral $ABCD$, $\hat{A} = \hat{B}$ and $BC = AD$. Prove that AB is parallel to DC.

10. If the opposite angles of a hexagon are equal, show that its opposite sides are parallel.
[See Exercise on p. 38.]

11. Prove that if the angles of a quadrilateral are in Arithmetical Progression, then the quadrilateral is a trapezium.

12. Prove that if the sides of a pentagon are produced so as to form a star-shaped figure, then the angles at the points of the star are together equal to two right angles. Extend this to any polygon having an odd number of sides.

13. From the vertex A of a right-angled triangle a perpendicular is drawn to the hypotenuse BC. The bisector of $\hat{B}$ meets this perpendicular in P and meets AC in Q. Prove that the triangle APQ is isosceles.

14. ABC is an isosceles triangle in which $AB = AC$ and $\hat{A} = 30°$. Points D and E are taken in BC (produced), AC respectively such that $A\hat{D}B = 55°$, $E\hat{B}C = 25°$. Calculate $B\hat{A}D$, $C\hat{A}D$ and prove that if AD, BE intersect at F, $AE = AF = BF$.

15. P, O, Q are three points in this order on a straight line. An acute angle POA is drawn on one side of PQ, and an angle $QOB = 3P\hat{O}A$ is drawn on the other side of PQ, OB being equal to OA. If AB cuts PQ at C, prove that $OC = CA$.

16. In the rider worked on p. 40, if the hexagon and the pentagon lie on opposite sides of AB, and CB is produced to meet XY at Q, prove that $QX = XC$.

17. In the figure on p. 40, calculate the magnitudes of the three angles of the following triangles:
   (a) that formed by AF, FE produced and AY produced;
   (b) that formed by AB produced in both directions,
       XY produced in both directions,
       and EF produced in both directions.

18. In a pentagon $ABCDE$, $AB = BC = EA$, the angles A, B are each 120°, and the angles C, E are each 95°. Alternate sides are produced to meet, forming a star-shaped figure. Calculate the magnitudes of the angles of the star.

19. $ABCDE$ is a convex pentagon with the side BC equal and parallel to AE and $AB = CD = DE$. If $\hat{A} = 70°$, find each of the other angles of the pentagon.

20. The external bisector of the vertical angle of a triangle makes an angle of 30° with the base. Prove that the difference of the base angles is equal to 60°.

21. If one pair of opposite angles of a quadrilateral are equal, prove that the bisectors of the other angles are parallel.

22. If *ABCDEFG* is a regular heptagon, calculate to the nearest minute the angles of the figure, the angles of the triangle *ACF*, and the angles of the triangle formed by *AC*, *DF* and the perpendicular at *G* to *FG*.

23. The sides *AB*, *DC* of a quadrilateral meet when produced at *P*, and the sides *BC*, *AD* meet when produced at *Q*. The lines bisecting the angles *P* and *Q* meet at *O*. Prove that $P\hat{O}Q$ or its supplement is equal to $\frac{1}{2}(B\hat{A}D + B\hat{C}D)$.

24. In a triangle *ABC*, $\hat{B} = 70°$ and $\hat{C} = 50°$; *P* is a point in *BC*, *Q* a point in *CA*, and *R* a point in *AB* such that $A\hat{P}C = 85°$, $A\hat{Q}B = 60°$, $B\hat{R}C = x°$. If *AP*, *BQ*, *CR* form a triangle *LMN*, prove that one angle of the triangle is of constant magnitude, and that the other two angles are equal if $x = 67\frac{1}{2}°$.

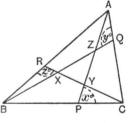

25. In the figure, $A\hat{B}C = 40°$ and $A\hat{C}B = 80°$, the other angles being as marked. Calculate the magnitudes of the angles of the triangle *XYZ*, and find the values of *y* and *z* if the triangle *XYZ* is equilateral and *AP* is perpendicular to *BC*.

CHAPTER VII

INEQUALITIES IN A TRIANGLE

The inequalities propositions may be classified as follows:

(i) *that having to do only with angles*—an exterior angle of a triangle is greater than either interior opposite angle of the triangle;

(ii) *that having to do only with sides*—any two sides of a triangle are together greater than the third side;

(iii) *those having to do with the relations between angles and sides*—the greater angle is opposite the greater side; and conversely.

It is usually possible to determine fairly easily to which of these three groups a rider on inequalities belongs. For instance, any exercises introducing a sum of sides would ultimately rest upon (ii). Again, if nothing is said about the lengths of the sides in a given figure in which the relative magnitudes of angles are discussed, it will be wise at any rate to start by applying (i). (iii), of course, will be used when both sides and angles are mentioned.

Other cases of inequalities may arise from the shortest distance of a point from a straight line, and from the properties of chords of a circle (see chapter XI).

Illustrative Riders

The angles ABC and ACB of a triangle ABC are bisected by BO and CO, which meet at O. If AB > AC, prove that OB > OC.

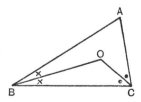

Ideas:

(i) Sides and angles (and not sides only; for bisected angles are mentioned) are referred to here.

(ii) We are told that $AB > AC$; make use of it.

We have $\qquad A\hat{C}B > A\hat{B}C.$

(iii) We are told that $\hat{B}$ and $\hat{C}$ are bisected; make use of it.

We have $\qquad O\hat{C}B > O\hat{B}C$, using (ii).

So $\qquad OB > OC.$

P is a point inside a triangle ABC. Prove that
$$AB + AC > PB + PC.$$

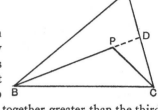

Ideas:

   (i) There is no mention
        of angles; we are only
        dealing with the sums
        of sides. We must
        therefore use — two
        sides of a triangle are together greater than the third
        side.

   (ii) $AB + AC > BC$, and also $PB + PC > BC$, but we
         cannot deduce from this that $AB + AC > PB + PC$.
         (Why?)

Some construction is therefore necessary.

   (iii) If we join AP we have
          $$AB + AP > PB, \quad AC + AP > PC.$$

This is no good for, on adding, we have
          $$AB + AC + 2AP > PB + PC.$$

   (iv) We are now left with producing BP to meet AC at D
         (or CP to meet AB).

   (v) We have first $AB + AD > BD$.

But we want AC, not AD; so add to each side DC.

We have $AB + AC > BD + DC.$

[This is a step forward, for $\triangle BPC$ bears to $\triangle BDC$ a like
relation as $\triangle BDC$ does to $\triangle ABC$.]

   (vi) Now $BD + DC = PB + PD + DC.$

[We want to get PB or PC on the right-hand side.]

But $PD + DC > PC.$

So all the more is $AB + AC > PB + PC.$

Prove that the perimeter of a quadrilateral is greater than the sum of its diagonals.

Ideas:

(i) This, again, only concerns the sum of sides, without reference to angles.

(ii) We have, in $\triangle ABC$,

$$AB + BC > AC,$$

and, in $\triangle ACD$, $CD + DA > AC$.

Adding these,

$$AB + BC + CD + DA > 2AC.$$

This is not quite what we want, as, though we have brought in the perimeter, we only have one diagonal.

(iii) However, the symmetry of our relation tells us that had we started with triangles containing BD we should similarly have obtained

$$AB + BC + CD + DA > 2BD.$$

So, adding and dividing right through by 2,

$$AB + BC + CD + DA > AC + BD.$$

We conclude with a rider which brings in the *median construction*.

Prove that any two sides of a triangle are together greater than twice the median to the third side.

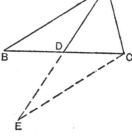

Ideas:

    (i) This again only concerns sides, not angles or angles and sides.

    (ii) As we saw on p. 14, any mention of medians always makes the medianal construction worth trying (unless Apollonius' Theorem is clearly indicated). This is especially the case here, where twice a median is mentioned.

So produce AD to E, making $DE = AD$, and join EC. As before, we prove the $\triangle$s ABD, CDE congruent.

    (iii) Here we are dealing only with sides, so $EC = AB$ is the fact arising from the congruence which we shall most likely require.

    (iv) Now we must bring in the inequality theorem, and $AE = 2AD$, so we have, in the $\triangle ACE$,

$$AC + CE > AE,$$

i.e. $$AC + AB > 2AD.$$

Exercise 6

1. P is any point inside a triangle ABC. Prove that
$$B\hat{P}C > B\hat{A}C.$$
[Angles only.]

2. Prove also that $AB + BC + CA > PA + PB + PC$ and that $PA + PB + PC > \frac{1}{2}(AB + BC + CA)$.
[See p. 46.]

3. Prove that the sum of three sides of a quadrilateral is greater than the fourth side.
[Draw a diagonal.]

4. Prove that the sum of the medians of a triangle is less than the sum of the sides, but greater than three-quarters of the sum of the sides.

5. D is the mid-point of the side BC of a triangle ABC. Prove that if $AD < BD$, the angle BAC is obtuse.
[Angle-sum, with inequalities.]

6. P is a point on the external bisector of the angle BAC of a triangle BAC. Prove that $PB + PC > AB + AC$.
[Produce BA to D, making $AD = AC$. Join PD.]

7. $ABCD$ is a parallelogram. If $AB > AD$, prove that
$$D\hat{A}C > B\hat{A}C.$$

8. In a triangle ABC, AD is drawn perpendicular to BC. If $AB > AC$, prove that $B\hat{A}D > D\hat{A}C$.

9. A and B are two fixed points on the same side of a given straight line XY. AC is drawn perpendicular to XY and produced to D so that $CD = AC$; BD meets XY at P. Prove that AP and BP are equally inclined to XY.

If Q is any other point in XY, prove that $AP + BP < AQ + BQ$.

10. Hence prove that of all the triangles on the same base and between the same parallels (i.e. which are equal in area) the isosceles triangle has the least perimeter.

11. In a parallelogram $ABCD$ the diagonal $BD >$ the diagonal AC. Prove that $\hat{A} > \hat{B}$.

12. P is a point inside a quadrilateral $ABCD$. Prove that $PA + PB + PC + PD$ is least when P is at the intersection of the diagonals.

13. If the angles of a parallelogram are not all equal, prove that the diagonal which joins the acute angles is longer than the other.

14. In a triangle ABC, $AB > AC$. If the bisector of $\hat{A}$ meets BC at P, prove that (a) $A\hat{P}B$ is obtuse; (b) $BP > PC$.

15. Prove that any two medians of a triangle are together greater than the third median.

[Use figure for proof of concurrence of medians.]

16. Prove that a diameter of a circle is greater than any other chord of the circle.

[Join the extremities of the chord to the centre.]

17. A straight line meets the circumferences of two non-intersecting circles at P, Q, R, S, in that order, and P, Q, R, S lie on the same side of the line of centres, which meets the circumferences at A, B, C, D, in that order. Prove that $AD > PS$, but that $BC < QR$.

[Use question 3 and the method of question 16.]

18. AB is a chord of a circle, centre O. Perpendiculars are drawn to AB from points in the major segment cut off by it. Prove that the greatest is the one which passes through O

[Draw a parallel through O to AB.]

CHAPTER VIII

THE AREA GROUP OF PROPOSITIONS

The fundamental proposition of this section is that *parallelograms on the same base and between the same parallels are equal in area*. From this we have that *triangles on the same base and between the same parallels are equal in area, and the converse;* and also expressions for the areas of parallelograms, triangles, quadrilaterals and trapezia.

There are many variations of the above which are of importance. Thus:

(i) Parallelograms, or triangles, on equal bases and between the same parallels (or of the same height) are equal in area; and conversely.

(ii) Triangles which have equal bases lying in the same straight line, and a common vertex, are equal in area.

(iii) If two triangles have a common vertex and their bases lie in the same straight line, their areas are proportional to the lengths of their bases. Similarly, mutatis mutandis, if they have equal altitudes.

When considering the areas of triangles, whether on the same base or not, it is frequently profitable to make use of the fact that their area is measured by half the product of a side and the altitude drawn to it. In particular, *equal triangles which have equal bases have equal altitudes*.

We have already seen that *parallelism* can be proved by the consideration of angles. The Area propositions are important in that they give us another test of parallelism; for equal triangles on the same base, or on equal bases in the same straight line, and on the same side of it, lie between the same parallels.

Note. *Equality of area must not be confused with congruence.* Plane figures which are congruent are also equal in area, but the converse does not necessarily hold.

Illustrative Riders

P is a point inside a parallelogram ABCD. Prove that $\triangle PAB + \triangle PCD = \frac{1}{2}ABCD$ in area.

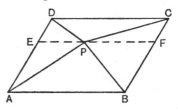

Ideas:

(i) What do we know about the areas of these triangles? They are not on bases in the same straight line, so all we can say (unless we use altitudes) is that their area is a half of any parallelogram on the same base and of the same height.

This suggests drawing EF through P parallel to AB.

(ii) We have then

$$\triangle PAB = \tfrac{1}{2} \text{ parm. } EFBA \text{ in area},$$

and $\triangle PCD = \tfrac{1}{2}$ parm. $EFCD$ in area;

hence adding, etc.

Using altitudes, since $AB = CD$, we should have

$\triangle PAB = \frac{1}{2}AB.p_1$, where p_1 is the perp. from P to AB,

$\triangle PCD = \frac{1}{2}CD.p_2$, where p_2 is the perp. from P to CD,

so $\triangle PAB + \triangle PCD = \frac{1}{2}AB(p_1 + p_2) = \frac{1}{2}$ parm. $ABCD$.

Prove that if each diagonal of a quadrilateral bisects its area, the quadrilateral is a parallelogram.

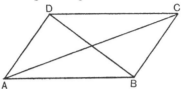

Ideas:

(i) This deals entirely with areas. We are only told that each diagonal divides the quadrilateral into two triangles of equal area.

(ii) What do we want to prove? That AB and DC, and that AD and BC, are parallel.

(iii) How can we connect this with areas? By proving the equality of area of two triangles lying between AB and DC (say), and on the same base (or on equal bases) and on the same side of it. $\triangle$s ADB, ACB suggest themselves for consideration.

(iv) What do we know of $\triangle ADB$? That it is $= \frac{1}{2}$ quad. $ABCD$.

Similarly	$\triangle ACB = \frac{1}{2}$ quad. $ABCD$.
So	$\triangle ADB = \triangle ACB$ in area,
i.e.	AB and DC are parallel.
Similarly	AD and BC are parallel.

Aliter (making use of altitudes).

Ideas:

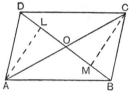

(i) $\triangle BCD = \triangle ABD$ in area, and they are on the same base. Draw the altitudes AL, CM.

Then $AL = CM$.

(ii) We wish to prove that $ABCD$ is a parallelogram. Can we say anything about (a) its opposite sides, (b) its opposite angles, or (c) that its diagonals bisect each other?

(iii) How can we best utilise $AL = CM$ in connection with (ii)? Remember we have AL is perpendicular to DO, and CM to OB.

For (a) or (b) we cannot prove such triangles as ADL, CMB or COD, AOB are congruent.

What of (c)?

In the $\triangle$s AOL, COM we already have

$$AL = CM, \quad A\hat{L}O = C\hat{M}O \text{ (right angles).}$$

So we must either have $LO = OM$ or another pair of angles equal.

Well, $A\hat{O}L = C\hat{O}M$ (vertically opposite).

So the triangles are congruent and $AO = OC$.

Similarly $DO = OB$, and the quadrilateral is therefore a parallelogram.

The area propositions give us very neat proofs of *the mid-point properties* of a triangle and of a trapezium.

Prove that the straight line joining the mid-points of two sides of a triangle is parallel to the third side.

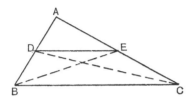

Ideas:

(i) What does $AD = DB$ give us in terms of areas? We have a triangle ADE on AD as base. This suggests joining BE. We then have

$$\triangle BDE = \triangle ADE \text{ in area.}$$

(ii) Similarly from $AE = EC$ we are led to join DC, and we have

$$\triangle CDE = \triangle ADE \text{ in area,}$$

whence $\triangle BDE = \triangle CDE$ in area,

and so DE and BC are parallel.

E and F are the mid-points of the oblique sides of a trapezium ABCD. Prove that EF is parallel to AB and CD.

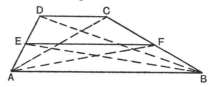

Ideas:

(i) *DC* is parallel to *AB*. How can we use this in relation to areas? By forming triangles on *AB* or *DC* as base—join *AC*, *BD*.

We have then △ *ADB* = △ *ACB* in area.

(ii) What do we want to prove? That *EF* is parallel to *AB*. How can we connect this with areas? By joining *EB*, *AF* (forming triangles on *AB* as base with *E*, *F* as vertices), and proving

△ *AEB* = △ *AFB* in area.

(iii) We have not yet made use of the fact that *E* and *F* are the mid-points of *AD* and *BC*. We have to bear in mind that we already have two triangles *ADB*, *ACB* equal in area, and that we wish to prove the triangles *AEB*, *AFB* equal in area.

The "area" interpretation of the fact that *E* and *F* are mid-points is therefore

△ *AEB* = $\frac{1}{2}$ △ *ADB* in area,
△ *AFB* = $\frac{1}{2}$ △ *ACB* in area.

(iv) So we have

△ *AEB* = △ *AFB* in area,

whence   *EF* and *AB* are parallel.

Aliter by altitudes.

Draw perpendiculars from *C*, *D*, *E*, *F* to *AB*. Those from *D* and *C* are equal. Using the mid-point propositions, prove that those from *E* and *F* are equal.

Exercise 7

1. *ABCD* is a parallelogram; *P* and *Q* are points in *BC*, *CD* respectively. Prove that $\triangle APD = \triangle AQB$ in area.

2. Prove that the diagonals of a parallelogram divide it into four triangles which are equal in area.

3. *D, E* are the mid-points of the sides *AB, AC* of a triangle *ABC*. Prove that $\triangle ADE = \frac{1}{4} \triangle ABC$ in area.

4. *O* is any point on the median *AD* of a triangle *ABC*. Prove that $\triangle OAB = \triangle OAC$ in area.

Hence find a point *P* within the triangle which is such that $\triangle PAB = \triangle PBC = \triangle PAC$ in area.

5. *E* is the mid-point of the diagonal *AC* of a quadrilateral *ABCD*. Prove that the quadrilateral *ABED* is equal to half the quadrilateral *ABCD* in area.

6. Prove that the straight line joining the vertices of two triangles of equal area which lie on the same base, but on opposite sides of it, is bisected by the common base.

[Altitudes.]

7. Prove that if a diagonal bisects the area of a quadrilateral, it bisects the other diagonal.

8. Prove, using areas, that the straight line drawn through the mid-point of a side of a triangle parallel to another side bisects the third side.

9. In question 12 of p. 27, prove that *RP* is parallel to *QS*.

10. If the mid-points of the sides of a quadrilateral are joined in order, prove that the area of the parallelogram so formed is one-half that of the quadrilateral.

[Consider the corner triangles.]

11. *AB* and *CD* are two straight lines meeting at *O*. If $\triangle AOD = \triangle COB$ in area, prove that *AC* is parallel to *BD*.

12. *D* is the mid-point of the side *BC* of a triangle *ABC*; *P* is any point in *DC*. *DQ* is drawn parallel to *AP* and meets *AB* at *Q*. Prove that the area of the quadrilateral *AQPC* is equal to a half that of the triangle *ABC*.

13. *ABCD* is a trapezium and *E* is the mid-point of *AD*, one of the oblique sides. Prove that the area of the triangle *EBC* is a half that of the trapezium.

14. *AB*, *XY*, *CD* are three parallel lines, *X* being on *AC* and *Y* on *BD*. Prove that the quadrilaterals *BXCY*, *AXDY* are equal in area.

15. *ABC* is a triangle. *D* is the point of trisection of *AC* nearer to *A*, and *E* is the point of trisection of *BD* nearer to *B*. What fraction of the area of the triangle *ABC* is the area of the triangle *BEC*? If *F* is the other point of trisection of *AC*, what fraction of the area of the triangle *ABC* is the area of the quadrilateral *BEFC*?

16. *P* is a given point and *ABCD* is a parallelogram. Show that the triangle *PAC* is equal to the sum or difference of the triangles *PAB*, *PAD* according as *P* lies outside or within $B\hat{A}D$ or its vertically opposite angle.

17. An equilateral triangle *ABD* is described on the side *AB* of a triangle *ABC*, right angled at *B*, *D* falling outside the triangle *ABC*; prove that the area of the triangle *DBC* is half the area of the triangle *ABC*.

[Draw *DE* parallel to *BC*, to meet *AB* at *E*.]

18. *ABCD* is a parallelogram; *BC* is bisected at *E*, and *CD* at *F*; *AE* and *DC* are produced to meet at *G*, and *AF* and *BC* are produced to meet at *H*. Prove that the area of the triangle *AGH* is equal to one and a half times the area of the parallelogram.

CHAPTER IX

PYTHAGORAS' AND ITS ASSOCIATED THEOREMS

The general enunciation of this group of theorems is that *in a triangle the square on any side is greater than, equal to, or less than the sum of the squares on the other two sides according as the angle opposite to the side is obtuse, right or acute; the difference in the first and last cases being equal to twice the rectangle contained by one of the two sides and the projection of the other upon it.*

Accordingly, to determine whether the greatest angle of a triangle is obtuse, right or acute, the best procedure is as follows:

    (i) Form the square on the greatest side (for the greatest angle must lie opposite to it).

    (ii) Form the sum of the squares on the other sides.

The triangle is obtuse, right or acute angled according as (i) is greater than, equal to, or less than (ii).

E.g. *Of what kind are the triangles whose sides are*

$$(a)\ 12'', 8'', 7'';$$
$$(b)\ 10'', 8'', 7''?$$

(a) $\qquad 12^2 = 144;\ 8^2 + 7^2 = 113.$

$\qquad\qquad$ ∴ the greatest angle is obtuse.

(b) $\qquad 10^2 = 100;\ 8^2 + 7^2 = 113.$

$\qquad\qquad$ ∴ the greatest angle is acute.

The *calculation of the projection* of one side upon another is frequently of importance.

E.g. *Find the area of the triangle whose sides are* 13″, 14″, 15″.

Here $15^2 = 225$; $13^2 + 14^2 = 365$;

∴ $\hat{C}$ is acute.

So we have, if AD is perpendicular to BC and $CD = x''$,

$$15^2 = 13^2 + 14^2 - 2 \cdot 14 \cdot x,$$

i.e. $28x = 140,$

$x = 5,$

so $AD^2 = 13^2 - 5^2 = 12^2,$

$AD = 12.$

Area of $\triangle ABC = \frac{1}{2}BC \cdot AD$

$= \frac{1}{2} \cdot 14 \cdot 12$

$= 84$ square inches.

We have already met right angles in various connections. The converse of Pythagoras' Theorem enables us either

(*a*) to test from the lengths of the sides of a triangle whether it is right angled, or

(*b*) to construct a right angle,

e.g. by drawing triangles whose sides are 3″, 4″, 5″; 5 cm., 12 cm., 13 cm.

It is useful to keep in mind the diagonal of a square when any reference to twice the square on a given line is mentioned.

Illustrative Riders

Some of the riders on Pythagoras' Theorem are obvious from their construction. Thus:

ABC is a triangle and P is a point within it. From P perpendiculars PD, PE, PF are drawn to BC, CA, AB respectively. Prove that
$$BD^2 + CE^2 + AF^2$$
$$= DC^2 + EA^2 + FB^2.$$

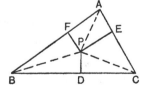

For to express BD^2 we must join BP, and the only thing we can say of BD^2 is that

$$BD^2 + PD^2 = BP^2,$$

or better, $\qquad\qquad BD^2 = BP^2 - PD^2.$

Similarly $\qquad\qquad CE^2 = CP^2 - PE^2,$

and $\qquad\qquad AF^2 = AP^2 - PF^2,$

so

$$BD^2 + CE^2 + AF^2 = AP^2 + BP^2 + CP^2 - PD^2 - PE^2 - PF^2.$$

The symmetry of our result shows us that we should similarly have

$$DC^2 + EA^2 + FB^2 = AP^2 + BP^2 + CP^2 - PD^2 - PE^2 - PF^2.$$

hence, etc.

Others require more complicated analysis, e.g.

Divide a straight line into two parts so that the difference of the squares on them is equal to a given square.

Ideas:

(i) We have to divide
    AB at C so that
    $AC^2 - CB^2 = k^2$,
    where k is given.

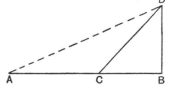

Write this in the form

$$AC^2 = CB^2 + k^2;$$

this is more suggestive of Pythagoras' Theorem.

(ii) $CB^2 + k^2$ suggests erecting a perpendicular BD equal to k (in order to form a right-angled triangle).

(iii) Join CD; then $CB^2 + k^2 = CD^2$.

(iv) We have then

$$AC^2 = CD^2 \quad \text{or} \quad AC = CD,$$

i.e. C is equidistant from A and D (gives perpendicular bisector).

(v) The required construction is therefore:

Draw BD perpendicular to AB and equal to k. Bisect AD at right angles. This bisector meets AB at the required point.

Further riders on the extensions of Pythagoras' Theorem for triangles having an angle of 60° or 120° are given on pp. 137, 138.

Apollonius' Theorem

If a rider deals with the sum of the squares on two *adjacent* sides of a figure, and Pythagoras' Theorem is not at once suggested (e.g. by mentioning only one other square, or by reference to right angles), Apollonius' Theorem is most probably indicated. Especially is this the case where mid-points are introduced or implied. We are also enabled, by this theorem, to calculate the lengths of the medians of a triangle when the sides are given.

We shall start with a numerical rider:

ABC is an equilateral triangle of side $2\sqrt{3}''$. At G, the intersection of the medians of the triangle, a straight line GP of length 4 units is erected perpendicular to the plane of the triangle. Calculate PA, PB, PC.

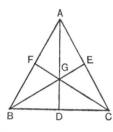 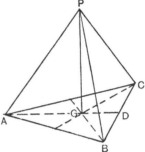

Ideas:

(i) We know that $AG = \frac{2}{3}AD$, so let us calculate AD.

(ii) Now $AB^2 + AC^2 = 2AD^2 + 2BD^2$,

i.e.
$$12 + 12 = 2AD^2 + 6,$$
$$AD^2 = 9,$$
$$AD = 3'',$$

so
$$AG = \frac{2}{3}.3 = 2''.$$

[We may also calculate AD from $\triangle ABD$ by Pythagoras' Theorem.]

(iii) Since PG is perpendicular to the plane of $\triangle ABC$,
$$P\hat{G}A = 1 \text{ rt. } \angle.$$

So, using Pythagoras' Theorem,
$$PA^2 = PG^2 + AG^2$$
$$= 16 + 4 = 20,$$

whence
$$PA = 2\sqrt{5}''.$$

Similarly, for $CG = AG = BG$, for an equilateral triangle,
$$PB = PC = 2\sqrt{5}''.$$

Prove that the sum of the squares on the sides of a parallelogram is equal to the sum of the squares on its diagonals.

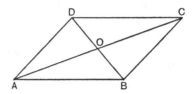

Ideas:

    (i) Sum of squares suggests Pythagoras' or Apollonius' Theorem. There are no right angles. This excludes Pythagoras' Theorem.

    (ii) Diagonals suggest: the diagonals of a parallelogram bisect one another (mid-points). This seems to confirm the suggestion of Apollonius' Theorem.

    (iii) So we have

$$AB^2 + BC^2 = 2AO^2 + 2BO^2,$$
$$CD^2 + DA^2 = 2AO^2 + 2DO^2.$$

Adding,

$$AB^2 + BC^2 + CD^2 + DA^2 = 4AO^2 + 2BO^2 + 2DO^2$$
$$= 4AO^2 + 4BO^2 \{DO = BO\}$$
$$= AC^2 + BD^2.$$

Note. If $AC = 2AO,$

               $AC^2 = 4AO^2,$ not $2AO^2,$

and if $BO = \frac{1}{2}BD,$

               $BO^2 = \frac{1}{4}BD^2,$ not $\frac{1}{2}BD^2.$

This is a very common error in Geometrical calculations.

Exercise 8

1. Prove that the square on the diagonal of a square is equal to twice the square on a side.

2. Prove that triangles whose sides are given by
$$m^2 + n^2, \quad m^2 - n^2 \text{ and } 2mn \text{ units}$$
are right angled for all positive values of m and n.

3. Determine whether the triangles whose sides are as follows are obtuse, right or acute angled:

$\quad$ (a) $2''$, $3''$, $\sqrt{6}''$;
$\quad$ (b) $4''$, $3''$, $\sqrt{7}''$;
$\quad$ (c) $\sqrt{3}''$, $\sqrt{5}''$, $\sqrt{13}''$.

Find the areas of the triangles.

4. P, Q are two points on a straight line perpendicular to a straight line AB, P being further from AB than Q. Prove that
$$PA^2 - QA^2 = PB^2 - QB^2.$$

5. Calculate the lengths of the medians of a triangle whose sides are $6''$, $8''$, $10''$ long.

6. The angle at A of a convex quadrilateral $ABCD$ is a right angle, and the diagonal BD is perpendicular to one of the sides. Show that the square described on the longest side of the quadrilateral is equal to the sum of the squares on the other three sides.

7. If the angles of a triangle are proportional to 1, 1, 2, show that the sides are proportional to 1, 1, $\sqrt{2}$.

8. Calculate to the nearest inch the length of the longest straight stick that can be packed inside a packing case of inside measurements 3 ft. by 4 ft. by 6 ft.

9. P is a point within a rectangle $ABCD$. Prove that
$$PA^2 + PC^2 = PB^2 + PD^2.$$

10. $ABCD$ is a rectangle in which AB is 16 cm. long and AD is 7 cm. long. In AB a point P is taken so that $AP = 4$ cm. Show that the angle subtended at P by DC is slightly less than a right angle.

11. Prove that the sum of the squares on the straight lines joining the mid-points of opposite sides of a quadrilateral is equal to half the sum of the squares on the diagonals of the quadrilateral.

12. The diagonals of a quadrilateral $ABCD$ are at right angles and $AB^2 + BC^2 = AD^2 + DC^2$. Prove that AC bisects BD.

13. If P is a point in the same plane as a parallelogram $ABCD$, prove that $PA^2 - PB^2 + PC^2 - PD^2$ is the same wherever P is, and that the expression is equal to half the difference of the squares on the diagonals of the parallelogram.

14. Prove that the sum of the squares on the sides of a quadrilateral exceeds the sum of the squares on the diagonals by four times the square on the straight line joining the mid-points of the diagonals.

15. Show that three times the sum of the squares on the sides of a triangle is equal to four times the sum of the squares on the medians.

16. A symmetrical pyramid stands on a square base $ABCD$ of side 8″ and each of its slant edges is 7″. V is its vertex and O the centre of the square base. Calculate its altitude VO which is at right angles to the base $ABCD$.

17. D, E, F are the mid-points of the sides AB, BC, CA of an equilateral triangle of side $4\sqrt{3}$. The triangle is folded about DE, EF, FD so as to form a regular tetrahedron, the vertices A, B, C all coinciding at a point O. Show that the height of the tetrahedron is $2\sqrt{2}$.

[Where will the perp. from O fall? What is the centre of $\triangle DEF$?]

18. ABC is a triangle having $AB = BC = 15$ cm., $AC = 18$ cm. A point O is taken on the bisector of $A\hat{B}C$ such that $BO = 8$ cm. At O a line OV, 15 cm. long, is drawn perpendicular to the plane of the triangle ABC. Employ Pythagoras' Theorem to calculate the lengths VA, VB, VC.

19. The mid-points of the sides BC, CA, AB of a triangle are D, E, F respectively; AD, BE, CF intersect at G. If $AD = 12, BE = 15, CF = 9$, prove that $D\hat{G}C$ is a right angle.

20. D is a point in the side BC of a triangle ABC such that $BD = 2DC$. Prove that

$$AB^2 + 2AC^2 = 3AD^2 + 6DC^2.$$

[Same method as Apollonius'.]

21. Divide a straight line into two parts so that the sum of the squares on them is equal to a given square.

[Isosceles right-angled triangle—$45°$.]

22. Divide a straight line into two parts so that the square on one part is double that on the other part.

23. If the sum of the squares on two opposite sides of a quadrilateral is equal to the sum of the squares on the other two sides, prove that the diagonals are at right angles.

[Drop perps. from two opposite vertices on a diagonal.]

24. G is the point of intersection of the medians of a triangle ABC. If $G\hat{C}B = 90°$, prove that $c^2 = 3a^2 + b^2$.

25. E and F are the points of trisection of the side BC of a triangle ABC. D is the mid-point of BC. Prove that

$$9\,(AE^2 + AF^2) = AB^2 + AC^2 + 16AD^2.$$

THE CIRCLE

CHAPTER X

ANGLE PROPERTIES OF A CIRCLE

The main property of this section is that which states that *the angle at the centre of a circle is double an angle at the circumference, standing upon the same arc*. The subsidiary facts about angles in a circle are:

(i) *Angles in the same segment are equal.*

(ii) *The opposite angles of a cyclic quadrilateral are together equal to two right angles.* Almost of equal importance is that *an exterior angle of a cyclic quadrilateral is equal to the interior opposite angle of the quadrilateral.*

(iii) *The angle in a semicircle is a right angle.*

Angles in the same segment must have a common base (which need not necessarily be drawn) and lie on the same side of it; and the vertices and the end points of the base must lie on the circumference. It is a useful (though not infallible) test to see whether angles quoted as equal *can* begin and end with the same pair of letters, e.g.

$$P$$

Angles AQB would be equal if A, B, P, Q, R lie on the

$$R$$

circumference of a circle and P, Q, R lie on the same side of AB.

Sometimes we make use of a set of concyclic points and do not draw the circle upon which they lie. In such a case it is important to see that when use is made of the properties of angles in the circle, the three points of each angle (e.g. in the angle BAC, B, A and C) are points which have been proved to lie on the circumference of the circle.

The converses of (i), (ii) and (iii) are very important as they give the conditions under which a circle will pass through a given set of points.

When a diameter is mentioned, and tangents are not introduced, (iii) will have to be used. If the centre is mentioned, the main property of the section will almost invariably be of use.

Illustrative Riders

If a parallelogram can be inscribed in a circle it must be a rectangle.

Ideas:

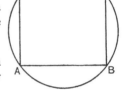

(i) A rectangle is a parallelogram with one angle a right angle (by definition).

(ii) We have a quadrilateral in a circle. What can we say of it?

$$\hat{B} + \hat{D} = 2 \text{ rt. } \angle s \text{ or } \hat{A} + \hat{C} = 2 \text{ rt. } \angle s.$$

(iii) We have a parallelogram. What are its properties?

(*a*) its opposite sides are equal—this is no good as we are dealing with angles;

(*b*) its opposite sides are parallel;

(*c*) its opposite angles are equal.

(iv) (*b*) would do as the idea of allied angles would lead us to a right angle; but (*c*) seems more obvious as it gives us another relation between $\hat{B}$ and $\hat{D}$.

We have $\hat{B} = \hat{D}$,

but $\hat{B} + \hat{D} = 2 \text{ rt. } \angle s \text{ (ii)}.$

So $\hat{B} = 1$ rt. $\angle$ and the parallelogram is therefore a rectangle.

ABCD is a cyclic quadrilateral. The bisector of the angle DAB meets the circumference at E. BC is produced to F. Prove that EC produced bisects the angle DCF.

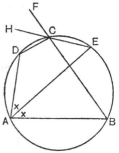

To prove

$$H\hat{C}F = H\hat{C}D \text{ or } H\hat{C}D = \tfrac{1}{2}F\hat{C}D.$$

Ideas:

I.   (i)   What can we say of $H\hat{C}D$ and $F\hat{C}D$? $H\hat{C}D$ is an exterior angle of the cyclic quadrilateral $AECD$.

So $H\hat{C}D = D\hat{A}E.$

[This seems a step in the right direction, for it brings us at once to a given fact, viz. $D\hat{A}E = E\hat{A}B$.]

     (ii)   Similarly $F\hat{C}D = D\hat{A}B$ (quadrilateral $ABCD$),

and $D\hat{A}E = \tfrac{1}{2} D\hat{A}B,$

so $H\hat{C}D = \tfrac{1}{2}F\hat{C}D.$

Or we might have proceeded:

II.   (i) $H\hat{C}D = D\hat{A}E$, as before.

    (ii) $H\hat{C}F$ is not an exterior angle of a cyclic quadrilateral. All we can say of it is that

          $H\hat{C}F = E\hat{C}B$, vertically opposite.

   (iii)   Now $E\hat{C}B$ has its three points E, C, B on the circumference.

So $E\hat{C}B = E\hat{A}B$

in the same segment. Hence, etc.

OA, OB are two radii of a circle such that the angle between them is 130°. *C is a point on the minor arc AB such that the angle OAC* = 45°. *Calculate the other angles of the quadrilateral OACB.*

Ideas:

(i) *AOB* and *ACB* are the only angles which have their feet on the circumference. They are therefore the only angles to which we can directly apply the angle properties of a circle.

(ii) Now $A\hat{C}B$ and the reflex $A\hat{O}B$ are two angles standing on the same arc *AB*, one at the centre and the other at the circumference, so we have

$$A\hat{C}B = \tfrac{1}{2} \text{ reflex } A\hat{O}B$$
$$= \tfrac{1}{2}(360° - 130°)$$
$$= 115°.$$

(iii) We now have three angles of the quadrilateral. The angle-sum property will give us the fourth.

AB, AC are the equal sides of an isosceles triangle inscribed in a circle. Any two points P, Q are taken in the base BC and AP, AQ produced meet the circumference of the circle at R and S respectively. Prove that P, Q, R, S are concyclic.

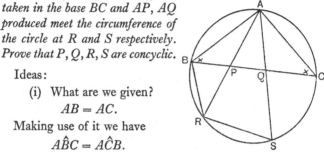

Ideas:

(i) What are we given?

$$AB = AC.$$

Making use of it we have

$$A\hat{B}C = A\hat{C}B.$$

(ii) How can we prove that *P, Q, R, S* are concyclic? From the figure it appears best to try to prove a pair of opposite angles supplementary. There is one other possibility—the rectangle property,

$$AP.AR = AQ.AS.$$

This will be dealt with later.

[It is no good joining *PS* and *QR*; a little reflection will show that the angles so formed cannot easily be expressed.]

(iii) Let us commence with an angle which we can easily express, $A\hat{S}R = 180° - A\hat{B}R$. [Join *BR*; then *ASRB* is cyclic.]

So we have to prove $A\hat{B}R = Q\hat{P}R$.

(iv) But $Q\hat{P}R = C\hat{B}R + B\hat{R}A$ (this way, for $C\hat{B}R$ is a part of $A\hat{B}R$).

So we have to prove $B\hat{R}A = A\hat{B}C$.

But $B\hat{R}A = B\hat{C}A$ (in the same segment)

                        $= A\hat{B}C$.   Hence, etc.

This is a good example of a rider in which "working from both ends" pays.

ABCD is a quadrilateral inscribed in a circle; the angle between BC and AD produced is 30° and the angle between AB and DC produced is 40°. Find the interior angles of ABCD.

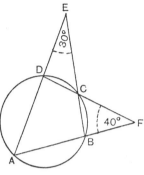

Ideas:

(i) *AEB* and *AFD* are not angles in a segment of a circle; nor can either be connected in this way with any angle in the circle circumscribing the quadrilateral.

(ii) We therefore fall back upon the angle-sum theorem, or its variation—exterior angle equal to sum of interior opposite angles.

(iii) Let $D\hat{A}B = x°$. [We choose it because it is a common angle of both the triangles *EAB*, *FAD*.]

(iv) We can now express all the angles of *ABCD* in terms of *x*.

For $\qquad D\hat{C}B = 180° - x,$

$\qquad\qquad C\hat{B}A = 180° - 30° - x = 150° - x \quad (\triangle ABE),$

$\qquad\qquad C\hat{D}A = 180° - 40° - x = 140° - x \quad (\triangle AFD).$

But $\qquad\qquad\qquad C\hat{B}A + C\hat{D}A = 180°.$

So $\qquad\qquad\qquad 150° - x + 140° - x = 180°,$

whence $\qquad\qquad\qquad x = 55°.$ Hence, etc.

Ex. Obtain this result by expressing

$$A\hat{E}B = C\hat{B}F - D\hat{A}B$$

and $\qquad\qquad\qquad A\hat{F}D = A\hat{B}C - B\hat{C}F.$

The student will see that there are many variations of this idea.

Two chords AB, CD of a circle intersect within the circle;
AC subtends an angle of 20° at the
centre, and BD subtends an angle of
30° at the centre. Find the angle
between the chords.

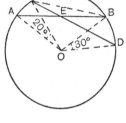

Ideas:

(i) The centre is directly mentioned; what does this suggest? Angles at the centre are double those at the circumference standing upon the same arcs.

So join *BC*. (*AD* would do just as well.)

(ii) We have $A\hat{B}C = \frac{1}{2}A\hat{O}C = 10°$,
and, in the same triangle,

$$B\hat{C}D = \frac{1}{2}B\hat{O}D = 15°.$$

(iii) How can we express $A\hat{E}C$ or $B\hat{E}C$? As in the last rider we are led to the angle-sum theorem. $A\hat{E}C$ is an exterior angle of the triangle *BEC*.

So $A\hat{E}C = E\hat{B}C + E\hat{C}B$

(we might have started with this)

$$= A\hat{B}C + D\hat{C}B$$
$$= 10° + 15°$$
$$= 25°.$$

OA is perpendicular to OB, and ON is perpendicular to AB. Points C, D are taken in OA, OB respectively such that CD is bisected by ON. Prove that ABDC is a cyclic quadrilateral.

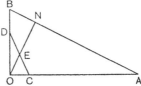

Ideas:

(i) *OA* is perpendicular to *OB*; *ON* is perpendicular to *AB*; *DE = EC*. The first and last facts suggest that *E* is the centre of the circle on *CD* as diameter, which passes through *O*.

So $$DE = OE = EC.$$

(ii) We want to prove that *ABDC* is a cyclic quadrilateral. As before, our figure suggests proving that

$$D\hat{C}A + D\hat{B}A = 180°,$$

or better that $$D\hat{C}O = D\hat{B}A$$

(for this brings us to the triangle *ODC*, about which we know something).

(iii) Can we evaluate $D\hat{C}O$ in any way? Yes, for we have

$$OE = EC.$$

This gives us $$D\hat{C}O = E\hat{O}C.$$

(iv) We have not yet used that *ON* is perpendicular to *AB*.

Well, $$E\hat{O}C \text{ (or } N\hat{O}A) = 90° - N\hat{A}O.$$

(v) We must bear in mind that we wish to prove

$$D\hat{C}O \text{ (i.e. } E\hat{O}C) = D\hat{B}A.$$

So we have to connect $D\hat{B}A$ and $N\hat{A}O$. But

$$D\hat{B}A = 90° - N\hat{A}O,$$

for *OA* is perpendicular to *OB*.

The proof may now be reconstructed.

The **altitudes of a triangle** are such a fruitful source of riders on angles in a circle that we add a short discussion of the six cyclic quadrilaterals associated with a triangle.

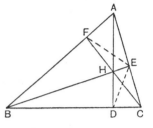

Draw the altitudes BE, CF. Let them meet at H. Join AH and produce it to meet BC at D. What have we? Right angles at E and F.

So $\qquad\qquad B\hat{F}C = 1 \text{ rt. } \angle = B\hat{E}C,$

i.e. $\qquad\qquad BCEF$ is a cyclic quadrilateral.

$\qquad\qquad$ [BC is a diameter of the circle.]

Also $\qquad\qquad A\hat{E}H + A\hat{F}H = 2 \text{ rt. } \angle\text{s},$

so $\qquad\qquad AEHF$ is a cyclic quadrilateral.

$\qquad\qquad$ [AH is a diameter of the circle.]

We shall now proceed to establish that

the three altitudes of a triangle are concurrent,

i.e. to prove that AD is perpendicular to BC.

This is equivalent to proving either $CDHE$ or $BDHF$ cyclic. (Why?)

Ideas:

    (i) We have to prove one of the following:

        (a) angles in the same segment equal;

        (b) a pair of opposite angles supplementary;

        (c) that $H\hat{D}C$ is a right angle, directly.

    (ii) Of the angles of $CDHE$ we know that $H\hat{E}C$ is a right angle and we also have a definite known angle in $\hat{C}$.

So can we prove $D\hat{H}E = 180° - \hat{C}$?

(iii) Let us try to express $D\hat{H}E$,

$$D\hat{H}E = 180° - A\hat{H}E.$$

But $AEHF$ is a cyclic quadrilateral, so

$$A\hat{H}E = A\hat{F}E,$$

and since $BCEF$ is a cyclic quadrilateral,

$$A\hat{F}E = \hat{C};$$

so
$$D\hat{H}E = 180° - \hat{C},$$

i.e. $CDHE$ is cyclic.

$$\therefore H\hat{D}C = 180° - H\hat{E}C$$
$$= 180° - 90°$$
$$= 90°.$$

The six cyclic quadrilaterals are

$$AEHF, BCEF;\ BDHF, CAFD;\ CEHD, ABDE.$$

Exercise 9

1. If one side of a cyclic quadrilateral is produced, prove that the exterior angle so formed is equal to the interior opposite angle of the quadrilateral.

2. Prove that the circles described on two of the sides of a triangle as diameters intersect on the third side, or on the third side produced.

3. Show that the circle drawn on one of the equal sides of an isosceles triangle as diameter passes through the middle point of the base; and that if the circle on the base as diameter meets the equal sides in P, Q, then PQ is parallel to the base.

4. [An exercise on "spotting" angles in the same segment.] Write down the angles

   (i) equal to

$$A (P_3 \hat{P}_1 P_5),\ B (P_2 \hat{P}_6 P_5),\ C (P_4 \hat{P}_2 P_6),$$
$$D (P_4 \hat{P}_3 P_5),\ E (P_5 \hat{P}_7 P_6);$$

   (ii) supplementary to A, B, C.

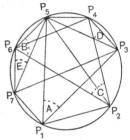

5. Two circles intersect at L and M. Two straight lines are drawn through L and M to meet the circumferences of the circles in A, B and D, C respectively. Prove that AD is parallel to BC.

6. $ABCD$ is a cyclic quadrilateral. If $\hat{A} = 2\hat{B}$, prove that
$$\hat{D} = \hat{B} + \hat{C}.$$

7. The altitudes AD, BE, CF of a triangle ABC meet at H. Prove that

   (a) $B\hat{H}C = 180° - \hat{A}$;

   (b) the altitudes are the internal bisectors, and the sides BC, CA, AB the external bisectors, of the angles of the triangle DEF (the pedal triangle).

8. If AD produced (see question 7) meets the circumcircle of the triangle ABC at P, prove that $HD = DP$; deduce that the circumcircles of the triangles ABC, BHC are equal.

9. If a kite can be inscribed in a circle, show that one of its diagonals must be a diameter of the circle.

10. The oblique sides of a trapezium inscribed in a circle meet when produced at an angle of $30°$. Calculate the angles of the trapezium.

11. The common chord of two intersecting circles subtends angles of $80°$ and $60°$ at their respective centres. Prove that any straight line drawn through one of their points of intersection and terminated by their circumferences subtends an angle of $110°$ at their other point of intersection. Give a generalisation of this rider.

12. A chord QP of a circle is produced to N so that PN is equal to the radius of the circle. N is joined to O, the centre of the circle, and NO is produced to T. Prove that

$$Q\hat{O}T = 3Q\hat{N}T.$$

13. $ABCD$ is a cyclic quadrilateral having its diagonals at right angles. Angle $A = 105°$ and angle $B = 93°$. Write down the size of the angles C and D. If angle $ACB = x°$, express the sizes of angles ACD and CDB in terms of x; and calculate the value of x.

14. APB, CPD are two chords of a circle which meet at right angles at P. Q is the middle point of AD. Prove that QP produced meets BC at right angles.

15. A chord AB of a circle subtends an angle of $30°$ at the centre O. Another chord BC is drawn on the side of OB remote from A so that $O\hat{B}C = 20°$; AD is drawn parallel to BC to meet the circle at D and CD is joined. Find the angles of the quadrilateral $ABCD$.

16. ABC is an acute-angled triangle in which $AB > AC$. P is a point in BC such that $AP = AC$. AP produced meets the circumcircle of ABC at Q. Prove that $BP = BQ$, and that if AM, BM are drawn parallel respectively to BC and AP to meet at M, then M lies on the circle.

17. The bisector of the interior angle B of a triangle ABC meets the bisector of the exterior angle C in a point E. The internal bisector of the angle A meets BE in D. Prove that the points A, D, C, E are concyclic.

18. $ABCD$ is a trapezium in which AB and DC are the parallel sides. AB subtends a right angle at C.

If $DC = AD = \frac{1}{2}AB$, prove that $\hat{D} = 2\hat{B}$.

19. $ABCD$ is a square, L and M the mid-points of BC and CD respectively. AM, DL meet at O. Prove that $ABLO$ is a cyclic quadrilateral. Prove also that, if BO is joined, $BO = AB$.

20. O is any point within a triangle ABC and P is any point in BC. The circle OPC meets AC at Q, and the circle OPB meets AB at R. Prove that a circle can be drawn through A, Q, O, R.

21. Circles are described on the sides AB, AC of a triangle ABC as diameters. The bisectors of the angle B meet the first circle in P and Q, and the bisectors of the angle C meet the second circle in P' and Q'. Prove that P, Q, P', Q' lie in a straight line.

[First prove that QP is parallel to BC.]

22. Equilateral triangles ABP, ACQ are described on the sides AB, AC of a triangle ABC, P and Q falling outside the triangle. BQ and CP intersect at R. Prove that A, R, C, Q are concyclic.

CHAPTER XI

THE PROPERTIES OF CHORDS
OF A CIRCLE

We shall divide the riders in this section into three classes:

(i) those dealing primarily with the mid-points of chords or with single chords;

(ii) those dealing primarily with the lengths of two or more chords;

(iii) those dealing primarily with chords through the points of intersection of two circles.

The fundamental facts of the section are that *the perpendicular from the centre of a circle upon a chord bisects it, and that the straight line joining the mid-point of a chord to the centre of a circle is perpendicular to that chord.* So, if a perpendicular from the centre of a circle upon a chord is mentioned, the fact that it bisects the chord must be used; and conversely, if a mid-point of a chord is mentioned, join it at once to the centre and make use of the fact that the join is perpendicular to the chord.

Numerical riders on the lengths of chords at given distances from the centre of a circle (and vice versa) at once fall under this heading, Pythagoras' Theorem furnishing the necessary connection.

We have already dealt with inequalities in chapter VII, p. 44; to what has already been said about inequalities of lengths we may now add (iv) the distances of chords from the centre of a single circle, and our matter for dealing with inequalities is complete.

The facts concerning the lengths of chords of a circle of course are: *equal chords of a circle are equidistant from the centre; and conversely. Of two unequal chords of a circle, the greater lies nearer the centre; and conversely.*

Illustrative Riders

A chord AB of a circle cuts a concentric circle at C and D. Prove that AC = DB.

Ideas:

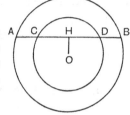

(i) We are dealing with lengths on a chord, and the centre is introduced (concentric circles), so we first drop a perpendicular *OH* on *AB* from the centre *O*.

(ii) This suggests mid-points, so we have

$$AH = HB \text{ from one circle,}$$

and $$CH = HD \text{ from the other.}$$

(iii) We have to prove $AC = DB$, so we merely subtract, obtaining $$AC = BD.$$

R is the mid-point of the common chord of two circles, centres A and B. Prove that ARB is a straight line.

Ideas:

(i) We are given a mid-point *R*, so join it to the centres *A* and *B*.

(ii) We have at once

$$P\hat{R}A = \text{a right angle,}$$
$$P\hat{R}B = \text{a right angle.}$$

(iii) What have we to prove? That A, R, B are collinear, i.e. that

$$P\hat{R}A + P\hat{R}B = 2 \text{ rt. } \angle\text{s};$$

and this follows from (ii).

N.B. It is useful to note that the problem of *calculating the length of the common chord of two intersecting circles*, the lengths of whose radii and line of centres are given, is the same as that which we meet in calculating the area of a triangle whose sides are given. (See page 60.)

For, referring to our last figure, since R is the mid-point of PQ, it means calculating the altitude PR of the triangle ABP, whose sides are given, and doubling it.

Chords of Intersecting Circles

Riders on chords through the intersection of two circles which deal with the *lengths* of the chords can generally be solved by the "mid-point" property.

N.B. Draw unequal circles unless you are expressly told that they are equal.

Of the straight lines drawn through a point of intersection of two circles, that which is parallel to their line of centres is the greatest.

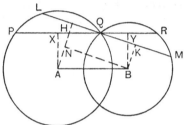

Ideas:

   (i) Draw perpendiculars $AX,\ BY$ and AH, BK upon the chords from the centres A, B

  (ii) Inequalities of lengths suggest:

        (a) the distances of chords from the centre of a (single) circle;

        (b) sides and angles of a triangle.

 (iii) (a) is not practicable, for though $AX < AH$, $BY > BK$. We are then left with (b).

 (iv) Perpendiculars give $PX = XQ,\ QY = YR$, so
$$PR = 2XY.$$
Similarly $\qquad\qquad\quad LM = 2HK.$
So we have to prove $\quad XY < HK.$

(v) We must use the fact that PR is parallel to AB. This gives us $XY = AB$ (for AX, BY are also parallel).

We must therefore prove $AB < HK$.

(vi) Can we get them into the same triangle? Well, if we draw BN parallel to HK to meet AH at N, $BKHN$ is a rectangle, so

$$HK = BN.$$

(vii) We have now to prove that in the triangle ABN,

$$BN < AB,$$

i.e. that $\qquad\qquad N\hat{A}B < A\hat{N}B.$

Do we know either of these angles? BN is parallel to HK, so

$$A\hat{N}B = A\hat{H}K = \text{I rt.} \angle.$$
$$\therefore N\hat{A}B < \text{I rt.} \angle$$
$$< A\hat{N}B.$$

Hence, etc.

As an example of a *numerical exercise* let us take:

The distance between two parallel chords of length 40 cm. and 48 cm. is 8 cm. The chords lie on the same side of the centre of the circle. What is the radius of the circle?

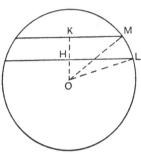

(i) Draw *OHK* perpendicular to the chords from the centre *O*. Let *r* cm. be the radius of the circle.

Then $KM = 20$ cm.; $HL = 24$ cm.; $HK = 8$ cm.

(ii) Using Pythagoras' Theorem we have, from triangle *OKM*,
$$(OH + 8)^2 + 20^2 = r^2,$$
and from triangle *OHL*,
$$OH^2 + 24^2 = r^2.$$

(iii) We now equate the two values of r^2; evaluate *OH*, and thence *r*.

We have
$$OH^2 + 16OH + 64 + 400 = OH^2 + 576,$$
$$16OH = 112,$$
$$OH = 7,$$
so
$$r^2 = 7^2 + 24^2 = 625,$$
$$r = 25 \text{ cm.}$$

Exercise 10

1. A chord 6″ in length is placed inside a circle of radius 5″. What is its distance from the centre?

2. Two circles, radii 20 cm. and 13 cm., intersect at two points which are 24 cm. apart. What is the distance between their centres?

3. Find the length of the common chord of two circles, radii 13″ and 15″, whose centres are 14″ apart.

4. Two parallel chords of a circle which lie on opposite sides of its centre are 10 cm. and 16 cm. in length and are 9 cm. apart. What is the radius of the circle?

5. A chord AB of a circle, radius 7″, meets a concentric circle of radius 5″ in C and D. If $AB = 10″$, find the length of CD.

6. Find the shortest chord that can be drawn through a given point within a circle.

7. AB, CD are two parallel chords of a circle, centre O; H is the mid-point of AB. If OH, produced if necessary, meets CD at K, prove that $CK = KD$.

8. H, K are the mid-points of two chords AB, CD of a circle, centre O. If $AB > CD$, prove that $O\hat{H}K > O\hat{K}H$.

9. Prove that if two chords are equally inclined to the diameter through their point of intersection, they are equal.

10. Two circles intersect at P and Q. A straight line parallel to PQ meets one of the circles at L and M, and the other at X and Y. Prove that $LX = MY$.

11. Prove that if a pair of parallel straight lines are drawn through the points of intersection of two circles, the portions of the straight lines intercepted between the circumferences are equal.

12. Draw the figure and work out the Exercise on p. 86 when LM cuts AB (not AB produced as in the given Exercise).

13. Two circles intersect at A and a chord is drawn through A meeting the larger circle at P and the smaller circle at Q so that $AQ = QP$. O, the centre of the larger circle, is joined to Q and produced to meet the smaller circle at R. Prove that

(i) AR is a diameter of the smaller circle;

(ii) $AR = PR$.

14. If two chords of a circle are equal to one another, show that they are equidistant from any point on the longest chord that can be drawn through their point of intersection.

15. Two straight lines are drawn through a point of intersection of two circles and terminated by their circumferences. If they are equally inclined to the common chord of the circles, prove that they are equal.

16. ABC is a triangle inscribed in a circle. Devise a construction for drawing a chord of the circle through A which shall be bisected by BC. Is this always possible?

17. If two intersecting chords of a circle are equal, then the segments of the one are respectively equal to the segments of the other.

18. Through P, one of the points of intersection of two circles, draw a straight line terminated by their circumferences such that it is bisected at P.

[Intercept theorem.]

19. From two points S, S' in the diameter of a semicircle, equidistant from the centre, are drawn parallel straight lines to meet the circumference in Y, Y'. Show that, if Y, Y' be joined, the angles at Y and Y' are right.

[Intercept theorem.]

CHAPTER XII

ARCS, ANGLES AND CHORDS

The facts concerning the relations of arcs, angles and chords are that in equal circles, or in the same circle (and this is often more important),

equal arcs cut off equal chords and subtend equal angles at the centre and at the circumference; and conversely, mutatis mutandis.

It follows that *if two arcs are in a certain ratio, the angles which they subtend, whether at the centre or at the circumference, are in the same ratio.* Thus, if one arc is double of another the angle which it subtends at the centre (or at the circumference) is double of the angle which the other arc subtends at the centre (or at the circumference). Again, if we are told that an arc is one-third (say) of a circumference, it subtends an angle of $\frac{1}{3}$ of 360°, or 120°, at the centre, and consequently an angle of 60° at the circumference.

The same thing, however, cannot be said of chords. Equal chords subtend equal angles, but it is quite wrong to say, for instance, that if one chord of a circle is double of another, the angle subtended by the one is double of that subtended by the other, whether at the centre or at the circumference; and conversely. If a figure is drawn, this is readily seen.

In general, *when arcs are mentioned they should at once be translated into terms of angles;* if this is of no help, then, if possible, in terms of chords.

In the riders of this section it is important, if possible, to fix upon two given points on the circumference of a circle, or upon a chord of given length, as we then have a basis for constant angles in the segments defined by them. In some cases the points, or chord, are given explicitly; in others, implicitly, as in two intersecting circles, where the points of intersection—or the common chord, which is of constant length—are frequently of great use as they determine segments in each circle which contain an angle which is constant in magnitude. Further examples of the use of this principle will be found in the section on Loci.

It is frequently helpful to note that *the perpendicular bisector of a chord bisects the arcs cut off by the chord*. This will be dealt with in the section on Symmetry.

Illustrative Riders

A and B are two fixed points on the circumference of a circle; P and Q are two variable points such that the arc PQ is of constant length. If PA, QB produced meet at R, prove that the angle ARB is constant.

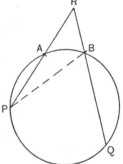

Ideas:

   (i) *A*, *B* are fixed points, so *AB* is of constant length. What does this mean in terms of angles? That $A\hat{P}B$ or $A\hat{Q}B$ is constant. Join *BP* (say).

   (ii) The arc *PQ* is of constant length. What does this mean in terms of angles? That $P\hat{A}Q$ or $P\hat{B}Q$ is of constant magnitude.

   (iii) So we have $P\hat{B}Q$ is constant and also $A\hat{P}B$ is constant.

   (iv) To express $A\hat{R}B$ we must use the angle-sum theorem. This gives us

$$A\hat{R}B = P\hat{B}Q - A\hat{P}B = \text{constant.}$$

Ex. Work out this rider, joining *AQ* instead of *BP*

Two circles intersect at P and Q; through these points straight

lines are drawn from a point R on the circumference of one of the circles to meet the circumference of the other circle at S and T. Prove that ST is of constant length for all positions of R.

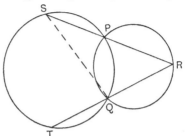

Ideas:

(i) We have to prove that ST is of constant length. In terms of angles this means to prove $S\hat{P}T$ or $S\hat{Q}T$ is constant. Join SQ (say).

(ii) S and T are not fixed points so it is no good considering $S\hat{Q}T$ as an angle in a segment.

How then can we express $S\hat{Q}T$?

(iii) We have, from the angle-sum property, that

$$S\hat{Q}T = R\hat{S}Q + S\hat{R}Q.$$

(iv) But $R\hat{S}Q$ (or $P\hat{S}Q$) is an angle in the segment cut off by PQ in the circle PSQ. And since PQ is fixed the angle PSQ is constant.

Similarly $P\hat{R}Q$ (or $S\hat{R}Q$) is constant in the circle PRQ. Hence, etc.

Ex. Work out this rider, joining PT instead of SQ.

If two chords of a circle intersect at right angles, the sum of each opposite pair of arcs cut off by them is equal to the semi-circumference.

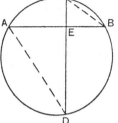

Ideas:

(i) Let us consider

arc AC + arc BD.

Translate into terms of angles at the circumference (for the centre is not mentioned).

(ii) (*a*) The angle subtended by a semi-circumference is a right angle;

(*b*) arc AC subtends either $C\hat{B}A$ or $C\hat{D}A$;

(*c*) arc BD subtends either $D\hat{C}B$ or $D\hat{A}B$.

Join CB, DA.

(iii) We are dealing with the sum of two angles, and that sum has to be a right angle. We must keep in mind that AB and CD are perpendicular; we have not used this yet.

The angles are not adjacent, so we cannot add them in that way. But taking together

$C\hat{B}A + D\hat{C}B$ or $C\hat{D}A + D\hat{A}B$ (in the same triangle),

in each case the third angle of the triangle is a right angle; hence, etc.

Ex. Work out this rider, joining AC and BD instead of BC and AD.

In an acute-angled triangle ABC the bisector of the angle A and the perpendicular from A on BC meet the circumcircle at L, M respectively. Prove that the arc LM is equal to half the difference of the arcs AB, AC. (Each is supposed to be less than half the circumference.)

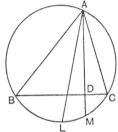

Ideas:

(i) We have to prove

arc $LM = \frac{1}{2}$ {arc $AB \sim$ arc AC}.

[In our figure $AB > AC$.]

(ii) Translate into terms of angles at the circumference. To prove $L\hat{A}M = \frac{1}{2}(A\hat{C}B - A\hat{B}C)$.

(iii) What are we given?

(a) $B\hat{A}L = C\hat{A}L$;

(b) AD is $\perp$ to BC. We must use both facts.

(iv) What will be our best way of expressing $A\hat{C}B - A\hat{B}C$? Using (iii) (b) we can get it all in terms of the angles at A, and then we should be able to bring in (iii) (a).

Thus $A\hat{C}B - A\hat{B}C = 90° - C\hat{A}D - (90° - B\hat{A}D)$

$= B\hat{A}D - C\hat{A}D.$

(v) Now, using (iii) (a),

$B\hat{A}D - C\hat{A}D = B\hat{A}L + L\hat{A}D - (C\hat{A}L - L\hat{A}D)$

$= 2L\hat{A}D.$

Hence, etc.

Prove that the straight lines which join the extremities of a pair of parallel chords of a circle towards the same parts are equal.

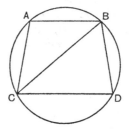

Ideas:

(i) We have to prove $AC = BD$. This is a rider on chords and angles (for parallels mean angles).

(ii) Translate $AC = BD$ into terms of angles. Join BC.
We have to prove $A\hat{B}C = B\hat{C}D$.

And these are alternate angles.

If A, B, C are three points in order on the circumference of a circle such that AB is a side of a regular 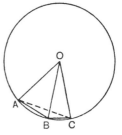 *10-sided polygon, and BC a side of a regular 15-sided polygon inscribed in the circle, prove that AC is equal to the radius of the circle.*

Ideas:

(i) The facts given us can only tell us something about the angles subtended at the centre or the circumference of the circle by *AB*, *BC*. Join *OA*, *OB*, *OC*.

We have
$$A\hat{O}B = \frac{360°}{10} = 36°,$$

$$B\hat{O}C = \frac{360°}{15} = 24°.$$

(ii) We have to prove $AC = OA = OC$, i.e. that the triangle *OAC* is equilateral.

But
$$A\hat{O}C = A\hat{O}B + B\hat{O}C$$
$$= 36° + 24°$$
$$= 60°.$$

Hence, etc.

Exercise 11

1. If two opposite sides of a quadrilateral inscribed in a circle are equal, prove that the other two sides are parallel.

2. *P* is any point on one of the arcs cut off by a fixed chord *AB* of a circle. Prove that the internal bisector of the angle *APB* cuts the conjugate arc in the same point for all positions of *P*. What can you say of the external bisector of the angle *APB*?

3. Devise a construction for drawing a triangle *ABC*, given *BC*, the angle *BAC* and the point where the internal bisector of the angle *BAC* meets *BC*.

4. Prove that if equiangular triangles are described in the same circle, or in equal circles, they must be congruent.

5. P is one of the points of intersection of two equal circles. Through P straight lines APB, CPD are drawn, terminated by the circumferences of the circles. Prove that $AC = BD$.

6. $ABCD$, APQ are a square and an equilateral triangle described in a circle. Prove that P is a point of trisection of the arc BC.

7. AB and AC are respectively sides of a square and a regular hexagon inscribed in a circle. Calculate, in degrees, the angle BAC.

8. Prove that the straight lines which join the extremities of a pair of parallel chords of a circle towards opposite parts are equal.

9. Through the points of intersection of two circles parallel straight lines are drawn, terminated by the circumferences. Prove that the straight lines which join their extremities towards the same parts are equal.

10. If a hexagon inscribed in a circle has two pairs of opposite sides parallel, prove that the third pair of sides also are parallel.

11. Two equal circles intersect at A and B. A straight line through A cuts the circles again at P and Q. Prove that

$$BP = BQ.$$

12. AB, BC are two sides of a regular 10-sided figure inscribed in a circle whose centre is O. AB, OC meet when produced at P. Prove that

    (i) $BP =$ a radius of the circle;

    (ii) $CP =$ a side of the figure;

    (iii) if BQ be drawn parallel to OC to cut the circle again at Q, then

    $BQ =$ a radius of the circle + a side of the figure.

13. Two adjacent sides of a quadrilateral inscribed in a circle subtend angles of 70° and 40° at the centre. The remaining two sides are equal. Calculate the angles of the quadrilateral.

14. If two equal chords of a circle intersect within the circle, prove that the segments of the one are equal to the segments of the other.

15. The perpendicular bisector of the side BC of a triangle ABC meets the circumcircle at D on the side remote from A; prove that AD bisects the angle BAC.

16. $ABCD$ is a cyclic quadrilateral; P, Q, R, S are the mid-points of the arcs AB, BC, CD, DA respectively. Prove that PR and QS are at right angles.
[Arc $QR = \frac{1}{2}$ arc BD.]

17. O is the centre of a circle and AB is an arc equal to one-ninth of the circumference. AC is drawn, outside the triangle AOB, so that the angle $OAC = 30°$, and C is on the circumference, CD is drawn parallel to OB, cutting the circle again at D. What fraction of the circumference is the minor arc BD?
[Translate.]

18. If two chords of a circle intersect within the circle they form an angle which is equal to the angle at the centre subtended by half the sum of the arcs which they cut off; and if they intersect outside the circle, that subtended at the centre by half the difference of the arcs which they cut off.

19. ABC is a triangle, right angled at A, and the bisector of the angle A cuts BC in D. If DF is drawn perpendicular to BC to meet AB in E and AC in F, prove that $DE = DC$, $DF = DB$ and $BE = CF$.

20. Two circles intersect at A and B; P and Q are points on the circumference of one of the circles such that the arc PQ is constant. PA, QB produced meet the other circle at L, M. Prove that the arc LM is constant.

CHAPTER XIII

TANGENCY

The fundamental proposition in this section is that

(i) *the tangent to a circle at a point is perpendicular to the radius drawn to that point.*

The other ideas to be kept in mind when tangents are mentioned are

(ii) *the two tangents drawn to a circle from an external point are equal;*

(iii) *the property of " the angle in the alternate segment".*

To " spot" angles in the alternate segment: suppose a straight line touches a circle and that through the point of contact a chord is drawn—then the alternate segment is that segment of the circle which lies on the other side of the chord from the angle. The circle need not be drawn, and in many cases actually is not drawn; for instance, we sometimes have to prove that points are concyclic by means of the converse of the theorem of the alternate segment.

Of rather less importance are that

(iv) *the straight line joining the centre of a circle to an external point bisects the angle between the tangents from that point; and also bisects the angle between the radii drawn to their points of contact.*

The most useful form of this is that if a circle touches two intersecting straight lines, its centre lies on a bisector of the angle between them.

Touching Circles

In this connection there are three things to be remembered:

(v) *They have a common tangent at their point of contact.* Draw it.

(vi) *Their centres and the point of contact are collinear.*

(vii) *The distance between their centres is equal to the sum or difference of their radii, according as the contact is external or internal respectively.*

As may be expected from (v) the property of the angle in the alternate segment is of frequent application with regard to riders concerning touching circles.

It is always important to note whether any mention (direct, or implied) is made of a centre or centres. If not, we can usually confine ourselves to (ii) and (iii) above.

Illustrative Riders

ABCD is a quadrilateral circumscribing a circle. Prove that
$$AB + CD = BC + AD.$$

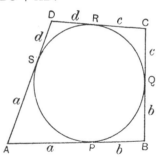

Ideas:

   (i) This question deals entirely with lengths, so we must start by writing down that the tangents drawn from each of the four vertices are equal.

I.e. $AP = AS$, $BP = BQ$, $CR = CQ$, $DR = DS$.

   (ii) Noting that $AB = AP + BP$, etc., our solution at once follows.

N.B. The solution of this and kindred problems is often simplified by attaching like letters (or symbols) to equal elements. For example, in the figure above we have
$$a + b + c + d = b + c + a + d.$$

Ex. What can we say if *ABCD* is (*a*) a parallelogram, (*b*) a rectangle?

Prove that two parallel tangents to a circle intercept on a third tangent a segment which subtends a right angle at the centre.

Ideas:

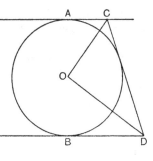

(i) We are not dealing with lengths, but with angles.

(ii) That $C\hat{O}D$ is a right angle does not seem capable of direct proof, so we are tempted to examine

$$O\hat{C}D + O\hat{D}C$$

(angle-sum property).

(iii) Parallel tangents suggest

    (*a*) alternate angles equal;

    (*b*) corresponding angles equal;

    (*c*) allied angles supplementary.

(*c*) seems to be the most likely as it suggests right angles.

(iv) Tangents suggest

    (*a*) perpendicular to radius (the centre is mentioned);

    (*b*) equal tangents from an external point;

    (*c*) angles between tangents bisected by OC and OD.

(*c*) brings in the allied angles, so let us connect it with (iii) (*c*).

(v) We have

$$O\hat{C}D + O\hat{D}C = \tfrac{1}{2}A\hat{C}D + \tfrac{1}{2}B\hat{D}C$$
$$= \tfrac{1}{2} \, 2 \text{ rt. } \angle\text{s}$$
$$= \text{I rt. } \angle.$$

Hence $\qquad\qquad C\hat{O}D = \text{I rt. } \angle.$

Two variable circles, centres P and Q, touch a fixed circle, centre A, internally and one another externally. Prove that the perimeter of the triangle APQ is constant.

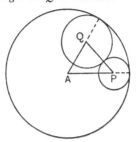

Ideas:

  (i) We are dealing with touching circles, and with lengths along their lines of centres. So property (vii) of p. 101 is clearly worth trying.

  (ii) Let us call the radii of the circles, centres A, P, Q, a, p, q respectively.

Then, using property (vii), we have

$$AP = a - p \quad \text{(internal touching)},$$
$$PQ = p + q \quad \text{(external touching)},$$
$$QA = a - q \quad \text{(internal touching)}.$$

So $\quad AP + PQ + QA = a - p + p + q + a - q$

$$= 2a = \text{constant}.$$

Two circles touch internally at P. Through P straight lines PAC, PBD are drawn which meet their circumferences at A, C and B, D. Prove that AB is parallel to CD.

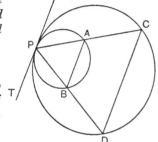

Ideas:

(i) Touching circles, no mention is made of centres; draw the common tangent *PT*.

(ii) If *AB* and *CD* are parallel we must have
  (*a*) alternate angles equal;
  (*b*) corresponding angles equal;
  (*c*) allied angles supplementary.

(iii) As no mention is made of centres we fall back on the alternate segment property, and start with that.

Let us take one of the circles first say *PCD*. We have

$$T\hat{P}D = P\hat{C}D.$$

(iv) Keeping (ii) in mind, we see (*b*) is the most convenient and we have, since *TP* touches the circle *PAB* also,

$$T\hat{P}B = P\hat{A}B.$$

So $$P\hat{A}B = P\hat{C}D,$$

i.e.   *AB* and *CD* are parallel.

Two circles intersect at A and B. P is any point on the circumference of one of the circles; PA and PB are joined and produced to meet the circumference of the other circle at C and D. Prove that the tangent at P is parallel to CD.

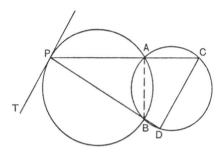

Ideas:

(i) No centre is mentioned; so the tangent suggests in this case the alternate segment property. Join AB.

We have $T\hat{P}B = P\hat{A}B.$

(ii) What have we to prove? That TP and CD are parallel.

We have already expressed $T\hat{P}B$, for we had in some way to make use of the fact that TP is a tangent. But if TP and CD are parallel,

$$T\hat{P}B = C\hat{D}P \qquad \text{(alternate)}.$$

So can we prove $P\hat{A}B = C\hat{D}P?$

(iii) But this is one of the first properties of the cyclic quadrilateral $ABDC$.

The proof may now be reconstructed.

ABCDE, ABHK are a regular pentagon and a square on the same base AB and on opposite sides of it; BE and HA produced meet at R. Show that CH touches the circle through B, R and H.

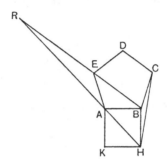

Ideas:

   (i) *CH* must *meet* the circle through *B, R, H* at *H*. Therefore, if it touches it at all, it must touch it at *H*.

   (ii) The centre of the circle is not mentioned or implied; we are not given two tangents; so we fall back upon the alternate segment property.

BH is the dividing line between *CH* and the circle; this shows us that we have to prove

$$B\hat{H}C = B\hat{R}H.$$

   (iii) As in the section on the angle-sum property (the most fruitful one when dealing with regular polygons) we have

$$B\hat{H}C = \tfrac{1}{2}(180° - H\hat{B}C) = \tfrac{1}{2}\{180° - (360° - 108° - 90°)\}$$
$$= 9°,$$

and $$B\hat{R}H = B\hat{A}H - E\hat{B}A = 45° - \tfrac{1}{2}(180° - E\hat{A}B)$$
$$= 45° - \tfrac{1}{2}(180° - 108°) = 9°;$$

hence, etc.

Exercise 12

1. Prove the rider on p. 104 for external contact.

2. Show that chords of a circle which are parallel to a tangent are bisected by the diameter drawn to the point of contact of the tangent.

3. If chords of the outer circle of two concentric circles touch the inner circle, prove that they are bisected at the point of contact.

4. P is the centre of a variable circle which touches two fixed circles, centres A and B, externally. Prove that $PA \sim PB$ is constant.

5. An ex-circle of a triangle ABC touches BC at D. Prove that
$$AB + BD = AC + CD.$$

6. The tangents from a fixed point A to a circle are met by a variable tangent in B and C. Prove that $AB + AC \pm BC$ is constant.

7. Two circles, centres A and B, touch externally at C. A straight line PCQ is drawn through C, terminated by their circumferences. Prove that (a) the radii AP, BQ; (b) the tangents at P and Q; are parallel.

8. Two circles intersect at A and B; LAM is a straight line through A terminated by their circumferences at L and M. The tangents at L and M meet at T. Show that L, T, M, B are concyclic.

9. Two circles touch externally at P. From any point on the common tangent at P tangents are drawn to the two circles. Prove that they are equal.

10. Using question 9, prove that a direct common tangent of two circles which touch externally subtends a right angle at their point of contact.

11. Prove that the straight line joining the centre of a circle to an external point bisects at right angles the chord of contact of the tangents drawn to the circle from that point.

12. ABC is a triangle in which $AB = AC$. A straight line through A meets BC at D and the circle through A, B and C in E. Prove that AB touches the circumcircle of the triangle BDE.

13. Prove that if a quadrilateral is described about a circle, the angles subtended at the centre by a pair of opposite sides are supplementary.

14. If a quadrilateral can be divided by a straight line crossing it so that circles can be inscribed in each of the two quadrilaterals into which it is divided, show that the length of the straight line is half the excess of the sum of one pair of opposite sides of the quadrilateral over the sum of the other pair.

15. Prove that the tangent to a circle at the mid-point of an arc is parallel to the chord of the arc.

16. The sides BC, CA, AB of a triangle ABC touch a circle at D, E, F. Calculate the magnitudes of the angles of the triangle DEF in terms of those of the triangle ABC.

17. Two circles intersect in the points P, Q and a common tangent touches the circles in the points A, B; prove that the angles APB, AQB are supplementary.

18. P is a point within an angle ABC, and is such that it is equidistant from A and B and also from AB and BC. Prove that BC touches the circle through P, A and B.

19. AP, AQ are two tangents to a circle whose centre is O. AO produced cuts the circle again in S. Prove that S is the centre of the circle which can be drawn to touch PQ, and AP, AQ both produced.

[Bisectors of angles.]

20. The inscribed circle of a triangle ABC meets BC, CA, AB at P, Q, R respectively. If $BC = 8''$, $CA = 4''$, $AB = 6''$, calculate the lengths of BP, PC, CQ, QA, AR and RB.

21. From a point P a tangent PA is drawn to a circle and a secant PBC cutting the circle at B and C. The bisector AQ of the angle BAC cuts BC at Q. Prove that $PQ = PA$.

[Translate to angles.]

22. OQ and OP are two perpendicular straight lines and OQ is equal to twice OP. A is the mid-point of OP and B is a point on OQ such that $OB = \frac{3}{8}OQ$. Prove that the circle with centre A and radius AP touches the circle with centre B and radius BQ.

23. Two circles touch at R and PQ is a direct common tangent; S and T are the extremities of the common diameter of the circles. SP and TQ meet at V. Prove that

(i) $VR = PQ$;

(ii) VR touches the circles at R.

[Diagonals of $VPRQ$?]

24. O is the mid-point of a line AB of length 12 cm. Semi-circles are described on OA, OB and AB. Calculate the radius of the circle which can be drawn to touch the smaller semicircles externally and the larger semicircle internally.

25. Two circles intersect at the points A, B and CD, a chord of one, touches the other at the point E. Prove that the angle $CAE =$ the angle EBD.

26. A circle is drawn to touch the sides BC, AB produced and AC produced of a triangle ABC. If I is the centre of this circle and AI meets the circumference of the circumcircle of the triangle ABC at P, prove that

$$PI = PB = PC.$$

27. The sides PQ, QR, RS, SP of a quadrilateral touch a circle at the points A, B, C, D respectively. Angle $QPS = 96°$ and angle $QRS = 120°$. AB, DC meet in E. Find by calculation the angle AED.

CHAPTER XIV

THE RECTANGLE PROPERTIES OF CIRCLES

The riders in this group may best be recognised if

(i) *the products of two segments of two different, intersecting, straight lines are mentioned.*

[*Note.* All four segments must have a common letter.]

As a sub-section of (i) we have

(ii) *A square equal to the rectangle contained by two segments of a straight line.*

[Here again the three lengths mentioned must have one common letter.]

This leads us to the "mean proportional" idea, associated with similar triangles. Indeed, the rectangle properties of a circle are intimately linked up with those of similar triangles.

In each case it is usually necessary to prove that a quadrilateral is cyclic. In all probability indications as regards angles will be given, or else facts with regard to sides or arcs which are translatable in terms of angles.

Note whether the intersection point (i.e. the common-letter point) is between or beyond the other two points in the same straight line.

If between, internal section is indicated. If a square is mentioned, look for a diameter.

If beyond, external section is indicated. If a square is mentioned, look for a tangent.

Frequently when we are given a square equal to a rectangle (the three segments having a common point not lying between two other points in the same straight line), there may be a connection between the rider and the alternate segment theorem. Thus a square equal to a rectangle relation may indicate to us that a certain straight line is a tangent to the circumcircle of the triangle formed by three points, and from that we may use the alternate segment theorem to obtain relations between angles (arcs or sides) of the figure (see question 16 in Exercise 13).

Conversely, the alternate segment theorem may enable us to prove that a certain straight line is a tangent, and thence a square equal to a rectangle relation will follow.

Illustrative Riders

ABC is a triangle right angled at A. AD is drawn perpendicular to BC. Prove that

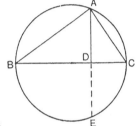

  I. $BD.DC = AD^2$.

  II. $BD.BC = AB^2$.

Ideas:

I. (i) The rectangle properties of a circle are suggested.

  $[BD.DC = AD.AD]$.

Note that the common point D lies between B and C; so the tangent form is not indicated.

(ii) Does the figure contain a cyclic quadrilateral of which BC is a chord? BC, because BD and DC lie in BC. No, but the angle BAC is a right angle, so the circle on BC as diameter passes through A. Draw it.

(iii) In order to bring in the rectangle $BD.DC$ we must have another chord through D. So produce AD to meet the circumference again at E.

Then $BD.DC = AD.DE$.

(iv) So we have to prove

$$AD.DE = AD^2,$$

i.e. $DE = AD$.

(v) Is this so? Yes, for BC is a diameter and AE is perpendicular to it.

The proof may now be reconstructed.

II. (i) Rectangular properties of a circle are again suggested, and in particular the tangent form, for in $BD.BC$, B, the common point of BD and BC, lies outside DC.

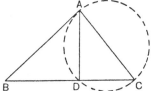

(ii) To prove
$$BA^2 = BD.BC$$
means that we have to prove that BA touches the circle through A, D, C.

(iii) But angle ADC is a right angle, so the circle through A, D, C has AC as its diameter; and BA is perpendicular to AC at A.

Hence, etc.

Note. It has already been remarked that where the rectangle properties of a circle may be used, there also are the *ratio theorems* likely to be of use. The riders just proved are a case in point.

For in I. it is only necessary to prove the triangles ABD, ADC similar, and in II. it is only necessary to prove the triangles ABD, ABC similar.

I. may be written $$\frac{BD}{AD} = \frac{AD}{DC}.$$

Notice that each side of this relation concerns three letters only, and that these letters denote the vertices of the triangles to be proved similar. This is an important method of noting the triangles which are to be proved similar. Similarly for II.

BE, CF are two altitudes of a triangle ABC. Prove that

$$AE.AC = AF.AB.$$

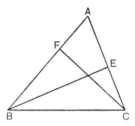

Ideas:

(i) The rectangle properties of a circle are suggested, for AE, AC and AF, AB each lie in straight lines. Moreover, since the straight lines AEC, AFB have a common point A, we see that it will be sufficient to prove that $BCEF$ is a cyclic quadrilateral.

(ii) What are we given? Altitudes.

How can we make use of them? Have we opposite angles of the quadrilateral supplementary, or angles in the same segment equal? Yes, for

$$B\hat{F}C = 1 \text{ rt. } \angle = B\hat{E}C.$$

$$\{\text{Note } B{\overset{F}{\underset{E}{\cdot}}}C\}$$

Hence, etc.

Ex. Obtain this result by means of similar triangles.

P is a point on the common chord AB produced of two inter-secting circles. PT is the tangent from P to one of the circles and a straight line through P meets the other circle at Q and R. Prove that the circle through Q, R and T touches PT at T.

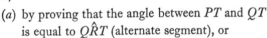

Ideas:

   (i) There are two methods of approach to this rider,

      (*a*) by proving that the angle between PT and QT is equal to $Q\hat{R}T$ (alternate segment), or

      (*b*) by means of the rectangle properties of a circle. This is suggested as PQR is *any* secant through P.

   (ii) (*a*) looks rather awkward, for $Q\hat{T}P$ does not seem directly expressible in terms of any convenient angle or angles. (Try this for yourself.)
     (*b*) means proving $PQ.PR = PT^2$.
     This seems more likely, for we can express both $PQ.PR$ and PT^2.

   (iii) We have in fact

$$PQ.PR = PA.PB \text{ (for circle } QRBA)$$

and $$PT^2 = PA.PB \text{ (for circle } ABT),$$

so $$PQ.PR = PT^2.$$

Hence, etc.

Numerical Examples

The height of a circular arc of radius r is h; the length of the chord of the arc is 2c. Prove that $c^2 = h(2r - h)$.

Ideas:

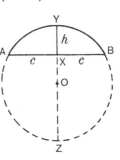

(i) The form $c.c = h(2r - h)$ (and the question itself) suggests the rectangle properties of a circle, so complete the circle. Let its centre be O, and the chord AB. Then if X is the mid-point of AB and XY is drawn perpendicular to AB, $XY = h$.

(ii) We can at once bring in c^2 and h by writing down

$$AX.XB = XY.(?)$$

Produce YX to meet the circumference at Z.

Then $$AX.XB = XY.XZ,$$

or $$c^2 = h.XZ.$$

(iii) We have thus to prove that

$$XZ = 2r - h,$$

and this is obvious from the figure.

Water flows in a cylindrical pipe whose radius is 1′ 8″. If the greatest width of the surface of the water in a right circular section is 2′, what is its greatest depth?

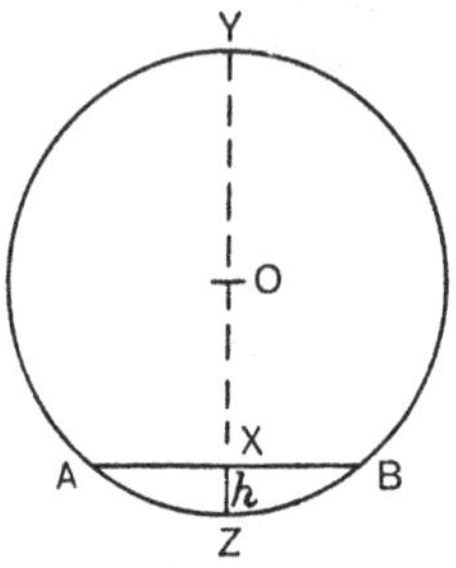

Ideas:

(i) Here again the rectangle properties of a circle are indicated (chord and height of arc).

In our figure we are given
$$AX = XB = 12″,$$
and YZ, the perpendicular diameter,
$$= 40″.$$
So we write down $AX.XB = XY.h,$

where $h = XZ$, the required depth.

(ii) This gives $12.12 = (40 − h).h$

for $XY = YZ − XZ,$

i.e. $h^2 − 40h + 144 = 0,$

$(h − 36)(h − 4) = 0,$

so $h = 36″$ or $4″.$

(iii) Why are there two solutions? The student should draw two figures and explain the two results.

Exercise 13

1. Prove that if two circles intersect, the tangents drawn to them from any point on their common chord produced are equal.

2. Prove that the common chord produced of two intersecting circles bisects their common tangents.

3. In the figure of the rider worked on p. 111 prove that
$$AC^2 = CB.CD;$$
and deduce that
$$BC^2 = AB^2 + AC^2.$$

4. Prove the first part of the rider on p. 111 by joining A to the mid-point of BC, and expressing $BD.DC$ as the difference of two squares.

5. If s is the shortest distance from an external point to a circle of radius r, and t is the length of the tangents from that point to the circle, prove that
$$t^2 = s (s + 2r).$$

6. Use question 5 to find roughly the diameter of the earth, if from a point 440 ft. above sea level the distance of the visible horizon is $44\frac{1}{2}$ miles.

7. The height of a circular arc of radius r is h; the length of the chord of half the arc is c. Prove that $c^2 = 2rh$.

8. From a point on the common chord of two intersecting circles secants are drawn to the circles, cutting them at A, B and C, D respectively. Prove that A, B, C, D are concyclic.

9. The tangent at A to a circle ABC meets the diameter BC at D, and AL is perpendicular to BC. Prove that
$$DB.DC - LB.LC = DL^2.$$

10. LM is a chord of a circle and it is bisected at K. DKJ is another chord. On DJ as diameter a semicircle is drawn and KS, perpendicular to DJ, meets this semicircle at S. Prove that $KS = KL$.

11. AB is the chord of contact of the tangents drawn to a circle centre O from a point P. If OP meets AB at Q, prove that $OP.OQ = r^2$, where r is the radius of the circle.

12. Prove the following construction for drawing a circle to touch a given straight line LM and to pass through two given points A, B on the same side of LM. Produce AB to meet LM at P. Draw any circle through A and B; let PX be a tangent to it from P. Measure PY along LM equal to PX. Then O, the point of intersection of the perpendicular at Y to LM with the perpendicular bisector of AB, is the centre of the required circle.

13. AB is a diameter of a circle, and AC a chord through A. Devise a construction for drawing a rectangle equal in area to the square on AC and having one side equal to AB.

14. M is the middle point of a line AB. Any line through A cuts the circle on AM as diameter in P and the line through B perpendicular to AB in Q. Prove that the rectangle $AP.AQ$ is equal to eight times the square on the radius of the circle.

15. A line BC touches a circle at C, and a straight line BAD cuts it in A and D. Show that the triangles ABC, CBD are equiangular, and express the lengths of DB, DC, DA in terms of the sides a, b, c of the triangle ABC.

[Similar triangles.]

16. ABC is an isosceles triangle having $AB = AC$ and each of these sides greater than BC. If a point K is taken in AB so that $BC^2 = BK.BA$, prove that the triangle BCK is isosceles.

17. Given a triangle ABC, in which AB is greater than AC, construct

(i) a point X on BC produced such that $XA^2 = XB.XC$, and

(ii) a point Z on BC itself such that $ZA^2 = ZB.ZC$, the triangle being such that this latter construction is possible.

18. $ABCDEFG$ is a regular heptagon. If a circle be drawn through F, D and X (the meet of GF and CD produced), prove that $CF^2 = CD.CX$.

19. If two chords of a circle intersect at right angles within the circle, prove that the sum of the squares on their segments is equal to the square on the diameter of the circle.

20. Two circles intersect at B and C and their direct common tangents AE, DF are drawn. If the common chord produced of the two circles meets the tangents at G and H, show that
$$GH^2 = AE^2 + BC^2.$$

21. Three circles intersect two by two. Prove that a point exists such that the tangents from it to all three circles are equal in length.

GENERAL

CHAPTER XV

LOCI

We do not propose to define loci and to discuss their general treatment. It is our purpose here merely to identify loci by a consideration of the ideas to which the forms of questions give rise.

It will not be out of place, however, to state that in order that a line, or lines (curved or straight), may be the locus which is sought two things must be established:

(i) *That every point which satisfies the given conditions must lie on the locus;* and

(ii) *That every point which lies on the locus must satisfy the given conditions.*

Thus, when we are asked to find the locus of points equidistant from two given points we first of all discover that each such point must lie on the perpendicular bisector of the straight line joining the given points, and then show that each point on the bisector is equidistant from the given points.

Again, the locus of a point which is equidistant from two given straight lines is the pair of straight lines which bisect the angles between them. One bisector only does not form the complete locus.

Certain propositions, from their very form, suggest the idea of a locus. Thus:

(i) *Triangles or parallelograms which lie on the same side of the same or equal bases and are equal in area lie between the same parallels.*

(ii) *The mid-point propositions,* for they also introduce the idea of *parallelism.*

(iii) *The converses of the propositions on angles in a circle,* e.g. that stating that angles in the same segment of a circle are equal.

(iv) *Apollonius' Theorem.* If the sum of the squares of two sides of a triangle which lies on a fixed base is given, the locus of the vertex is a circle.

We must not forget that most important of loci—*the circumference of a circle.* If we can establish that a variable point is at a constant distance from a point which is fixed in position, the locus of the variable point is a circle whose centre is at the fixed point. See pp. 123, 130.

Difficulty is often experienced by a student in defining accurately the segment of a circle which comprises a locus when an angle has been proved to be of constant magnitude. The safest procedure is—look down each arm of the angle which has been proved constant in size. Then if there is a fixed point on each of the arms (say A and B) the segment of the circle has AB as base.

The manner of attacking a locus problem depends a good deal upon the figure. It is a good plan to get a rough idea of the shape of the locus by drawing in a few positions of the point whose locus is sought. But this is not always possible when the figure for just one position is fairly complicated. Examples of both methods are given in what follows. ·

Illustrative Riders

Find the locus of the point of intersection of the diagonals of a parallelogram which lies on a fixed base and is of constant area.

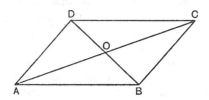

Ideas:

(i) Fixed base, constant area, suggest parallelograms or triangles on the same base and between the same parallels; in other words, an area question seems to be indicated.

(ii) O is a vertex of the $\triangle\ OAB$ which has a fixed base AB. What about its area?

(iii) Well, we know that $\triangle\ OAB = \frac{1}{4}$ parm. $ABCD$ in area. Its area is therefore constant; hence, etc.

Ex. Work out this question, making use of altitudes.

A rod slides with its ends upon two other rods fixed at right angles to each other. Find the locus of its mid-point.

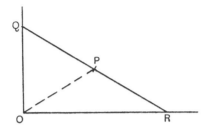

Ideas:

(i) What are the facts mentioned? (*a*) QOR is a right angle; (*b*) P is the mid-point of QR; (*c*) QR, being a rod, is of fixed length.

(ii) What do (*a*) and (*b*) suggest? The circle on QR as diameter, centre P, passes through O.

[*Note.* This is not a fixed circle, so we have not yet found the locus.]

(iii) How can we bring in (*c*)? Using (ii),

$$OP = PQ = PR = \tfrac{1}{2}QR = \text{constant};$$

and O is a fixed point.

The required locus is therefore a circle, centre O, radius equal to half the length of the rod.

O is a fixed point. A point P moves on a fixed straight line.
Find the locus of the mid-point of OP.

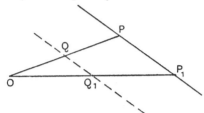

Ideas:

(i) Our figure gives us very little to work on. Let us take
another position of P and Q, say P_1, Q_1.

[Note the suffix notation, a useful one in problems of this
kind.]

(ii) Now Q and Q_1 are both mid-points. This suggests
the mid-point propositions.

We have at once that QQ_1 is parallel to PP_1. QQ_1 is therefore
a straight line fixed in direction.

(iii) Is it fixed in position? We can say vaguely that it lies
midway between O and PP_1. Can we fix it more
precisely?

Yes, for there must be a point P_2 which is the foot of the
perpendicular from O to PP_1. The corresponding Q_2 must be
in the same line as QQ_1. But Q_2 is a fixed point. Hence, etc.

Find the locus of mid-points of
    I. *parallel chords;*
    II. *chords through a fixed*
        point; of a circle.

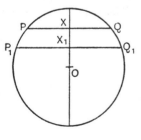

Ideas:

I.   (i) The mid-point of a chord
suggests joining it to the
centre of the circle. Then
$O\hat{X}P$ is a right angle.

Thus OX is fixed in direction (for it is perpendicular to PQ, which is fixed in direction); and it passes through a fixed point O. So, as OX is a fixed straight line, we will test it to see if it is the required locus.

(ii) Let it meet a parallel chord P_1Q_1 at X_1. Parallels suggest

      (a) alternate angles equal;

      (b) corresponding angles equal;

      (c) allied angles supplementary.

(iii) Either of these facts tells us that OX_1 is perpendicular to P_1Q_1. So X_1 is the mid-point of P_1Q_1.

The diameter perpendicular to the parallel chords is therefore the locus.

II. (i) As before, a mid-point suggests joining it to the centre of the circle. Then

      $O\hat{X}A$ is a right angle.

(ii) PQ is not fixed in direction, so we must consider the right angle in a different light.

(iii) OX passes through the fixed point O, and XA through A, i.e. X lies on the circle on OA as diameter— a fixed circle.

The reader should now test this for the locus by considering another chord through A.

Through points on the circumference of a given circle parallels of constant length are drawn in a fixed direction in the same sense. Prove that the locus of their extremities is an equal circle.

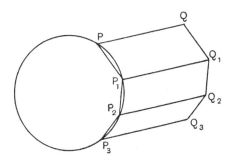

Ideas:

    (i) Here, again, a single position of P does not give us much to work on. If we draw several positions we have some idea of the form of the locus.

    (ii) What do equal and parallel straight lines suggest? Parallelograms.

    (iii) What is the condition that the points P, P_1, P_2, P_3 are concyclic?

Use the properties of parallelograms to obtain a like condition for Q, Q_1, Q_2, Q_3.

The new centre may similarly be obtained.

The reader should work out this question completely.

Two circles intersect at A and B; P is a variable point on one of the circles. PA, PB meet the circumference of the other circle at Q and R. Find the locus of X, the point of intersection of BQ and AR.

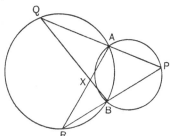

Ideas:

    (i) Nothing specific is given in this question, such as mid-points, right angles, etc. But we have A and B, fixed points on the circumferences of both circles. We must bear in mind that the angles subtended by AB at the circumferences of both circles are constant, i.e. that $A\hat{Q}B = A\hat{R}B =$ constant, and $A\hat{P}B =$ constant.

    (ii) X is the vertex of an angle whose arms pass through the fixed points A and B. Can we evaluate $A\hat{X}B$?

    (iii) We have $A\hat{X}B = A\hat{Q}B + Q\hat{A}R$

(or $A\hat{R}B + Q\hat{B}R$, which will ultimately come to the same thing).

Now $A\hat{Q}B$ is constant (see (i)).

Let us then examine $Q\hat{A}R$.

    (iv) We get no further by saying that $Q\hat{A}R = Q\hat{B}R$ (why?).

But $Q\hat{A}R = A\hat{R}P + A\hat{P}R$,

and both $A\hat{R}P$ and $A\hat{P}R$ are constant (see (i)).

So $A\hat{X}B$ is of constant magnitude and the locus of X is a segment of a circle whose chord is AB.

Owing to the importance of this method we give another example:

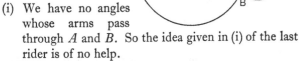

A is one of the points of intersection of two circles. Straight lines CAD, PAQ are drawn terminated by the circumferences. CAD is fixed, PAQ is variable. PC and DQ meet at X. Find the locus of X.

Ideas:

(i) We have no angles whose arms pass through A and B. So the idea given in (i) of the last rider is of no help.

(ii) What are we given? That CAD is fixed. We must use this. Translating it into terms of angles we have

$C\hat{P}A$ is of constant magnitude,

$A\hat{Q}D$ is of constant magnitude.

(iii) X is the vertex of an angle whose arms always pass through the fixed points C and D. Can we evaluate $C\hat{X}D$?

(iv) We have either

    (a) $C\hat{X}D = P\hat{Q}D - X\hat{P}Q$

or   (b) $C\hat{X}D = 180° - X\hat{P}Q - X\hat{Q}P$.

[Note how we have made it possible to introduce $C\hat{P}A$ and $A\hat{Q}D$, about which we know something.]

(v) Using (ii) we see that (a) tells us that $C\hat{X}D$ is of constant magnitude, (b) gives us the same result when we note that $A\hat{Q}D = 180° - X\hat{Q}P$.

The locus of X is therefore a segment of a circle whose chord is CD.

A and B are two fixed points on the circumference of a circle. PQ is an arc of constant length. Find the locus of the point of intersection of AP and BQ.

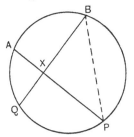

Ideas:

    (i) We are given two things:

        (*a*) *A*, *B* are fixed points;

        (*b*) arc *PQ* is of constant length.

(*a*) translates into a constant angle at the centre or the circumference, i.e.

$$A\hat{P}B \quad \text{or} \quad A\hat{Q}B.$$

(*b*) Similarly from *PQ* we have that

$$Q\hat{B}P \text{ and } Q\hat{A}P \text{ are constant.}$$

[The centre is not mentioned or implied in any way.]

    (ii) *X* is the vertex of an angle whose arms pass through the fixed points *A* and *B*. Let us then try to evaluate $A\hat{X}B$.

We have $A\hat{X}B = A\hat{P}B + Q\hat{B}P$

                      = a constant, from (i).

The locus of *X* is therefore a segment of a circle whose chord is *AB*.

Ex. Obtain a like result (i) by joining *AQ*, (ii) when *AP, BQ* intersect outside the circle.

Find the locus of the extremity of a tangent of given length to a circle.

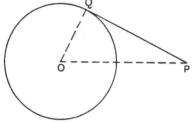

Ideas:

(i) A single tangent from P; we must therefore use

$$O\hat{Q}P = 1 \text{ rt. } \angle.$$

(ii) We have OQ of fixed length, PQ of constant length, i.e. we are given two sides of the right-angled triangle OQP.

So we have, using Pythagoras' Theorem,

$$OP^2 = OQ^2 + PQ^2 = \text{constant}.$$

$\therefore OP$ is of constant length; and O is a fixed point.

The locus of P is therefore a circle concentric with the given circle.

Aliter.

(i) We are given so little, i.e. such a simple figure, that we may draw another position of PQ, say P_1Q_1.

Then we have $\qquad PQ = P_1Q_1$,

$$\text{rt. } \angle OQP = \text{rt. } \angle OQ_1P_1.$$

(ii) Can we prove the triangles OQP, OQ_1P_1 congruent? We must have another angle, or another side (right angles).

But $OQ = OQ_1$; so the triangles are congruent and

$$OP_1 = OP.$$

Hence, etc.

A and B are fixed points. Find the locus of a point P which moves so that $AP^2 + BP^2$ is constant.

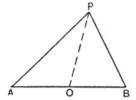

Ideas:

(i) What does $AP^2 + BP^2$ suggest? (*AP* and *BP* are not collinear)

    (*a*) Pythagoras' Theorem,

or   (*b*) Apollonius' Theorem.

(ii) It cannot mean (*a*), for $A\hat{P}B$ is not necessarily a right angle. So we try (*b*).

Let *O* be the mid-point of *AB*.

Join *OP*.

(iii) Then
$$AP^2 + BP^2 = 2AO^2 + 2OP^2.$$

So $AO^2 + OP^2$ is constant.

But $AO = \frac{1}{2}AB$ and is therefore constant.

So OP^2, and hence *OP*, is constant; and *O* is a fixed point (the mid-point of *AB*).

The locus of *P* is therefore a circle, centre *O*.

P is a point within a circle of radius 5 cm., QR is any chord through P. If QP . PR = 9 sq. cm., find the locus of P.

Ideas:

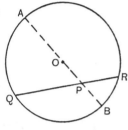

(i) The rectangle properties of a circle are suggested.

(ii) For any *one* position of *P*, the direction of *QR* is immaterial (for the rectangle contained by the segments will always be equal to 9 sq. cm.).

Let us therefore consider a chord through *P* which only depends upon the position of *P*, i.e. the diameter *AB* which passes through *P*.

(iii) We have

$$AP . PB = QP . PR$$

$$= 9 \text{ sq. cm.}$$

But

$$AO = OB = 5 \text{ cm.,}$$

so

$$(5 + OP)(5 - OP) = 9,$$

$$25 - OP^2 = 9,$$

i.e.

$$OP = 4 \text{ cm.}$$

The locus of *P* is therefore a circle, centre *O*, radius 4 cm.

A and B are two fixed points. Find the locus of the point of contact of the tangents from a point C on AB produced to all the circles through A and B.

Ideas:

    (i) The figure suggests the rectangle properties of a circle. Note that C, A, B are fixed points.

    (ii) We have

$$CP^2 = CA . CB \quad \text{(for } CP \text{ is a tangent)}$$
$$= \text{constant.}$$

The locus of P is therefore a circle, centre C, radius $\sqrt{CA.CB}$.

Aliter.

Here again we might have taken more than one circle through A and B. Then we should have had

$$CP^2 = CA.CB$$
$$= CP_1^2,$$

whence $\quad CP = CP_1$, etc.

Exercise 14

Find the locus of

    1. The mid-points of equal chords of a circle.

    2. The centres of circles touching two given straight lines.

    3. The centres of circles touching a given straight line at a given point.

    4. The centres of circles touching two parallel straight lines.

    5. The centres of circles touching a given straight line, and having a given radius.

    6. The centres of circles touching a given circle at a given point.

7. The centres of circles touching a given circle, and having a given radius.

8. The centres of circles touching two concentric circles.

9. The centres of circles passing through two given points.

10. The points of contact of tangents drawn from a fixed point to a system of concentric circles.

11. The points of contact of tangents drawn to a system of circles, which touch one another at a given point, from a point on their common tangent.

12. The point of intersection of tangents to a given circle which contain a given angle.

13. The intersection of medians of a triangle which has a fixed base and is of constant area.

14. ABC is a triangle lying on a fixed base BC, and having a constant vertical angle BAC. BA is produced to D so that $AD = AC$. Find the locus of D.

15. In question 14, if AP is cut off from AB and made equal to AC, find the locus of P.

16. A and B are fixed points. Find the locus of a point P which moves so that $AP^2 - PB^2$ is constant.

17. A and B are fixed points on the circumference of a circle; PQ is a variable diameter. Find the locus of the point of intersection of PA and QB.

18. P is a point outside a circle, radius 5 cm. A straight line through P meets the circumference of the circle at Q and R. If $PQ.PR = 24$ sq. cm., find the locus of P.

19. $ABCD$ is a rectangle. Find the locus of a point P which moves so that $PA^2 + PB^2 + PC^2 + PD^2$ is constant.

20. Given the middle points of two adjacent sides of a rectangle, prove that the loci of its vertices and of the intersection of its diagonals are all circles.

21. *ABCD* is a quadrilateral. Find the locus of a point *P* which moves so that the area of *ABPD* is always half the area of *ABCD*.

22. Through two fixed points *A*, *B* on the circumference of a circle any pair of parallel lines are drawn, cutting the circle in points *P*, *Q*. Prove that the line *PQ* in all its positions touches a second circle concentric with the given circle.

23. *P* is a point outside a circle of which *LM* is a diameter. *PM* meets the circle again in *N*. If *P* moves in such a way that the rectangle *PN*.*PM* is constant, prove that the locus of *P* is a circle and find its centre. Find also the locus of the middle point of *PL*.

24. *ABC* is a triangle; find the locus of a point *P* such that $P\hat{A}B = P\hat{C}A$, *P* lying inside the triangle.

25. *A* is a fixed point lying outside a fixed straight line *BC*. *P* is a variable point in *BC*. If a point *Q* is taken in *AP* so that *AP*.*AQ* is constant, prove that the locus of *Q* is a circle passing through *A*.

26. *A* is an extremity of a diameter of a circle. In *AP*, a chord of the circle, a point *Q* is taken such that *AP*.*AQ* is constant. Prove that as *P* varies, the locus of *Q* is a straight line perpendicular to the given diameter.

CHAPTER XVI

ANGLES OF 60° AND 120°

The relations between the sides of triangles containing an angle of 60° or 120° are often very interesting. The fundamental fact in this section is that, considering *an equilateral triangle* as a special isosceles triangle, *the bisector of the vertical angle bisects the base at right angles, thus dividing it into two equal parts each of which is equal to half the side of the triangle*.

Find the area of an equilateral triangle of side a.

Ideas:

(i) Draw AD, the bisector of the angle BAC. Then AD bisects BC at right angles and $BD = \dfrac{a}{2}$.

(ii) We have to find $\frac{1}{2}AD.BC$, i.e. $\frac{1}{2}AD.a$. So we must evaluate AD.

(iii) Now $AB = a$, $BD = \dfrac{a}{2}$ and $A\hat{D}B = 90°$, so, using Pythagoras' Theorem,

$$AD^2 = AB^2 - BD^2$$
$$= a^2 - \frac{a^2}{4} = \frac{3a^2}{4},$$

whence
$$AD = \frac{a\sqrt{3}}{2},$$

and the area of the triangle $ABC = \dfrac{a^2\sqrt{3}}{4}$.

A triangle ABC has $\hat{C} = 90°$, $\hat{B} = 30°$. *Prove that*
$$AB = 2AC.$$

Ideas:

   (i) We must use the idea of an equilateral triangle, so complete the equilateral triangle of which the triangle *ABC* forms a part.

   (ii) To do this, produce *AC* to *D* making $CD = AC$. Then the triangles *ABC*, *DBC* are congruent. (Why?)

Hence $B\hat{D}C = B\hat{A}C$
$$= 60°,$$

and so the triangle *ABD* is equilateral. (Why?)

   (iii) We have then
$$AB = AD = 2AC.$$

In a triangle ABC, $\hat{C} = 60°$. *Prove that*
$$AB^2 = AC^2 + BC^2 - BC.AC.$$

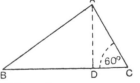

Ideas:

   (i) The form of the question suggests an extension of Pythagoras' Theorem.

So, drawing *AD* perpendicular to *BC*, we have
$$AB^2 = AC^2 + BC^2 - 2BC.CD.$$

   (ii) Comparing this with what we have to prove we see that we have to show that
$$CD = \tfrac{1}{2}AC.$$

   (iii) $\hat{C} = 60°$, so complete the equilateral triangle of which *ACD* forms a half.

The proof is left to the reader.

Exercise 15

1. In a triangle PQR, $\hat{R} = 90°$, $\hat{Q} = 60°$. Prove that
$$QR = \tfrac{1}{2}PQ.$$

2. In a triangle ABC, $\hat{C} = 120°$. Prove that
$$AB^2 = AC^2 + BC^2 + AC.BC.$$

3. Find the area of a regular hexagon of side $2''$.

4. If the angles of a triangle are proportional to $1 : 2 : 3$, prove that the sides are proportional to $1 : 2 : \sqrt{3}$.

5. ABC is an equilateral triangle inscribed in a circle, centre O. OD is drawn perpendicular to BC and produced to meet the circumference at E. Prove that BC bisects OE at right angles.

6. Use question 5 to prove that if a is the side of an equilateral triangle inscribed in a circle of radius r, then
$$a^2 = 3r^2.$$

7. Show that if a is the side of an equilateral triangle, and b that of a square, both of which are inscribed in the same circle, then
$$2a^2 = 3b^2.$$

8. ABC is an equilateral triangle, D is a point in BC produced. Prove that
$$AD^2 = BC^2 + CD^2 + BC.CD.$$

9. Two tangents to a circle of radius $1''$ include an angle of $60°$. Calculate their length.

10. Calculate the radius of the circle inscribed in an equilateral triangle of side 10 cm.

11. In a triangle ABC, $AB = 7''$, $AC = 8''$ and $\hat{A} = 60°$. Calculate the length of the median through A.

12. AB is a straight line divided internally at any point C. On AC, CB equilateral triangles APC, CQB are described on the same side of AB. Prove that
$$PQ^2 = AB^2 - 3AC.BC.$$

CHAPTER XVII

SYMMETRY

There are three forms of symmetry:

(*a*) Point symmetry.

(*b*) Line symmetry.

(*c*) Plane symmetry.

Point symmetry. When a figure (not necessarily a plane figure) has point symmetry, any straight line which joins opposite corresponding points and passes through a point of symmetry is bisected at that point; and, in the case of closed plane figures like parallelograms, equally divides the area of the figure.

Examples are found in:

A parallelogram with regard to the intersection of its diagonals.

A circle with regard to its centre.

A sphere with regard to its centre.

Line symmetry. A figure is said to be symmetrical about a line when one side of it can be folded through two right angles about that line as axis so as to coincide exactly with the other side of the figure. This definition also applies to solid figures. We have an elementary illustration in the case of an isosceles triangle. The bisector of its vertical angle is an axis of symmetry for the figure.

In connection with line symmetry it is useful to note that a straight line joining corresponding points with regard to a line (or axis) of symmetry is bisected at right angles by the line of symmetry.

Plane symmetry. This is an extension of line symmetry. A plane of symmetry divides a figure into two identical, interchangeable portions.

Examples are:

Any diametrical plane of a sphere.

Any plane through the vertex of a right circular cone, perpendicular to the base of the cone.

It is of the greatest importance to note that considerations of symmetry do not furnish us with a rigorous proof of the existence of any particular property or properties of a figure. Symmetry may often help us by indicating what is necessary to be proved, or what lines a particular proof may follow. And, of course, what we have rigorously proved for one side of a symmetrical figure may be taken as valid for the other side.

The reader should now work through the following examples, noting any points or line of symmetry; and also determining, from considerations of symmetry, what results might be expected to be capable of rigorous proof.

1. Isosceles triangle.　　　2. Equilateral triangle.

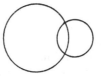

3. (*a*) Two intersecting circles; (*b*) and the common chord; (*c*) and any chords through the intersections.

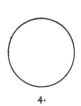 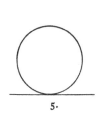 

4.   5.   6.

4. (*a*) A circle; (*b*) and one chord; (*c*) and two chords.

5. (*a*) A circle and tangent; (*b*) a circle and tangent with a chord parallel to the tangent.

6. A circle and two tangents. (Consider the chord of contact.)

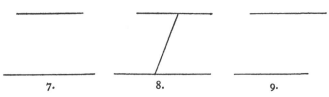

7.   8.   9.

7. Two parallel straight lines.

8. Two parallel straight lines and a terminated transversal.

9. Two equal and parallel straight lines.

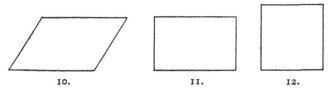

10.   11.   12.

10. A parallelogram.   11. A rectangle.

12. A square.

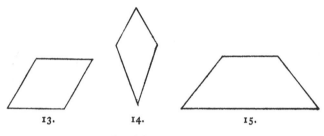

13. A rhombus.
14. A kite.
15. A trapezium with equal obliques.

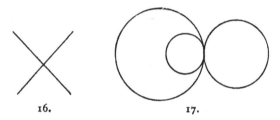

16. A pair of intersecting straight lines.
17. Two or more touching circles.

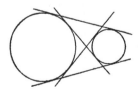

18. Two circles and their common tangents.

Afterwards, the following riders should similarly be considered:

Chapter II, p. 7.

Exercise 1, question 14.

Chapter III, pp. 12, 16.

Exercise 2, questions 1, 2, 3, 5–8, 14–17, 21–23.

Chapter IV, pp. 21, 22.

Exercise 3, questions 1, 2, 13, 17.

Exercise 4, questions 4, 13, 16, 18.

Exercise 5, questions 7, 9.

Exercise 9, questions 3, 9, 10.

Chapter XI, pp. 84, 85.

Exercise 10, questions 6, 7, 9, 10, 14, 15, 17.

Chapter XII, p. 96.

Exercise 11, questions 1, 8, 14.

Exercise 12, questions 3, 11, 15.

Chapter XV, pp. 124, 125.

Exercise 14, questions 1, 2, 3, 4, 6, 8, 9, 19.

Printed in the United States
By Bookmasters